The Stokes and Navier–Stokes Equations in Exterior Domains: Moving Domains and Decay Properties

Vom Fachbereich Mathematik

der Technischen Universität Darmstadt

zur Erlangung des Grades eines

Doktors der Naturwissenschaften

(Dr. rer. nat.)

genehmigte

Dissertation

von

David Wegmann, M. Sc.

aus Seeheim-Jugenheim

Referent: Prof. Dr. Reinhard Farwig

Korreferent: Prof. Dr. Mads Kyed

Tag der Einreichung: 18. Oktober 2018

Tag der mündlichen Prüfung: 07. Dezember 2018

Darmstadt 2019

D17

Bibliografische Information der Deutschen Bibliothek

Die Deutsche Nationalbibliothek verzeichnet diese Publikation in der
Deutschen Nationalbibliografie; detaillierte bibliografische Daten sind
im Internet über http://dnb.d-nb.de abrufbar.

ISBN 978-3-8325-4839-1
ISSN 1862-5681
zugl.: Darmstadt, Technische Universität Darmstadt, Dissertation

Logos Verlag Berlin GmbH
Comeniushof, Gubener Str. 47,
10243 Berlin
Tel.: +49 030 42 85 10 90
Fax: +49 030 42 85 10 92
INTERNET: http://www.logos-verlag.de

Contents

Preface

In the beginning of the nineteenth century, C.-L. Navier proposed a system of equations to model the motion of an incompressible viscous Newtonian fluid which is still the most famous model in the theory of fluid dynamics. The first derivation of these equations has been published in the middle of the same century by G. G. Stokes. Since then, the study of this system is one of the main topics in applied science. To honour Navier and Stokes, the system was named after them.

The *Navier–Stokes equations* with so-called *Dirichlet boundary conditions* are given by

$$\begin{aligned}
u_t - \nu\Delta u + u \cdot \nabla u + \nabla p &= f && \text{in } \Omega \times (0,T), \\
\operatorname{div} u &= 0 && \text{in } \Omega \times (0,T), \\
u &= \beta && \text{on } \partial\Omega \times (0,T), \\
u(0) &= u_0 && \text{on } \Omega.
\end{aligned}$$

(NST)

Here and hereafter, $\Omega \subset \mathbb{R}^3$ denotes a bounded or unbounded domain and $T \in (0,\infty]$. The unknown are the velocity field u of the fluid and the corresponding pressure p. By scaling arguments one can always assume that the viscosity $\nu = 1$. Furthermore, f denotes the external force, u_0 the initial value of the velocity field, and β the (inhomogeneous) Dirichlet boundary value.

The analytical treatment of (NST) mostly deals with the questions of existence and uniqueness of solutions, and blow-up phenomena. The answer to these question strongly depends on the notion of solutions to (NST).

The first breakthrough in the mathematical theory of the Navier–Stokes equations has been developed by J. Leray in the 1930's [35,36] and evolved by E. Hopf [31]. The existence of a global in time weak solution to the Navier–Stokes equations was already proved by J. Leray and the corresponding solution space $L^\infty(0, T; L^2_\sigma(\Omega)) \cap L^2(0, T, \dot{H}^1_0(\Omega))$ is still called Leray–Hopf class to honour their pioneering work. Note that the integrability of these so-called Leray–Hopf type weak solutions does not ensure that we can insert the solution itself as a test-function in the variational formulation. This lack of integrability is possibly the main reason why it is still an open problem whether Leray–Hopf type weak solutions are unique or not. In fact, it was proven by J. Serrin [47] following the ideas of J. Leray that Leray–Hopf type weak solutions are unique if the solution fulfils an additional integrability condition, *i.e.*, if the weak solution u fulfils

$$\text{(SC)} \qquad u \in L^s(0, T; L^q(\Omega)), \text{ where } \frac{2}{s} + \frac{3}{q} = 1.$$

Furthermore, it is also known, that if there exists a weak solution with this additional integrability condition (SC), then all Leray–Hopf type weak solutions which fulfil the strong energy inequality coincide. In contrast to the global in time existence of Leray–Hopf type weak solutions, however, the existence of a weak solution with the integrability condition (SC) is just known under an additional smallness condition, *i.e.*, if the data u_0 and f are assumed to be sufficiently small, or if T is assumed to be small.

A similar result as for Leray–Hopf type weak solutions with the integrability condition (SC) is known for strong solution, *i.e.*, for solutions u such that u_t and the second derivatives of u are bounded in a Lebesgue–Bochner space. Also, the existence of strong solutions is known for a large class of domains and these solutions are unique; as above, for the proof of existence one needs to assume a smallness condition.

Let us also mention, that recently, T. Buckmaster and V. Vicol proved that in the class of weak solutions, which are not bounded in the Leray–Hopf class, the uniqueness of solutions is not fulfilled. Here, a weak solution fulfils the Navier–Stokes equations in a distributional sense, but we do not assume any conditions on derivatives of the solution.

Even though the mathematical treatment of the Navier–Stokes equations began more than eighty years ago, the question of uniqueness and well-posedness is still an open problem and became one of the famous Millennium Prize Problems.

The most common way to construct a solution to the Navier–Stokes equations is to split the space of solutions into the solenoidal subspace and its complement. By applying the Helmholtz projection, the projection onto the solenoidal subspace, the Laplacian becomes the so-called Stokes operator

$$A_\Omega := -P_\Omega \Delta,$$

where P_Ω denotes the Helmholtz projection on the set Ω. Then the Navier–Stokes equations are transformed to

$$u_t + A_\Omega u + P_\Omega(u \cdot \nabla u) = P_\Omega f \quad \text{in } \Omega \times (0, T),$$
$$u(0) = u_0 \quad \text{on } \Omega.$$

Note that the incompressibility condition $\operatorname{div} u = 0$ and the boundary value is hidden in the domain of the Stokes operator.

In many applications it is not reasonable to assume that the domain Ω is fixed. For example, if one considers a falling stone in a liquid filled vessel, then the domain which is filled with liquid changes. In this example there might be a strong mutual influence of the sinking velocity of the stone and the movement of the fluid. Nevertheless, if the influence of the velocity of the fluid on the evolution of the domain is low, then it seems to be reasonable to consider as a first approximation the Navier–Stokes equations in a time-dependent domain

$$Q := \bigcup_{t>0} \Big(\Omega(t) \times \{t\} \Big).$$

In the first part of this dissertation we prove that the Navier–Stokes equations possesses a unique strong solution in the non-cylindrical time-space domain Q. Here, we assume that for each t the domain $\Omega(t)$ is an exterior domain with sufficiently smooth boundary. A solution u is called strong solution to the Navier–Stokes equations if

$$\partial_t u, A_{\Omega(\cdot)} u \in \mathrm{L}^s(0, \infty; \mathrm{L}^q_\sigma(\Omega(t))), \quad 1 \leq s, q < \infty,$$

and u fulfils the differential equations almost everywhere.

A strategy to solve such a non-linear system of differential equations is to prove a so-called *maximal L^s-regularity* estimate to the linearised problem, *i.e.*, we consider the Stokes equations

$$u_t + A_{\Omega(t)}u = Pf \quad \text{in } Q,$$
$$u(0) = u_0 \quad \text{on } \Omega(0)$$

first and prove that the solution operator is continuously invertible if we endow the space of solutions $\mathbb{E}$ with the norm

$$\|u\|_{\mathbb{E}} := \|\partial_t u\|_{L^s(0,\infty;L^q(\Omega(t)))} + \|A_{\Omega(t)}u\|_{L^s(0,\infty;L^q(\Omega(t)))}.$$

The first step to solve the Stokes equations in the non-cylindrical time-space domain is to transform the system onto a cylindrical reference domain. Therefore, we introduce suitable isomorphisms $\Phi(t)$, $0 \leq t < T$, which map functions defined on $\Omega(0)$ onto functions defined on $\Omega(t)$ and hence this transformations will allow us to consider the Stokes equations on the space $\Omega(0) \times (0,T)$. The transformation of the time derivative and the Stokes operator on the domain $\Omega(t)$ leads us to a system with non-autonomous operator since

$$\Phi(t)\partial_t\Phi(t)^{-1} =: \partial_t + B(t),$$
$$\Phi(t)A_{\Omega(t)}\Phi(t)^{-1} =: \tilde{A}(t)$$

with time-dependent operators $B(t)$ and $\tilde{A}(t)$, and thus the proof of a maximal regularity estimate to the Stokes equations on Q can be performed by proving that the non-autonomous system

$$u_t + (B(t) + \tilde{A}(t))u = Pf \quad \text{in } \Omega(0) \times (0,\infty),$$
$$u(0) = u_0 \quad \text{on } \Omega(0)$$

has maximal L^s-regularity. The proof of this maximal L^s-regularity result for the non-autonomous Stokes system will be presented in Chapter 2 and this will be one of the main results of this dissertation.

Once we have this strong result for the linear problem on hand, the proof of an existence theorem for the Navier–Stokes equations is relatively simple. The crucial part for this proof is to estimate the non-linear term by the norm of $\mathbb{E}$. Then, by assuming a suitable smallness condition on the data, the non-linear term is subordinate to the linear terms. Then the proof of existence can be done by Banach's Fixed Point Principle. The necessary smallness condition can be easily achieved by restricting to results locally in time. This has been done by J. Saal [46] more than 10 years ago. More crucial is a result if the time interval equals $[0, \infty)$ and the data, in particular, f and u_0, are assumed to be sufficiently small. We will consider this in Chapter 3.

As discussed above, the Navier–Stokes system we solve in Chapter 3 is a model for a fluid in a domain which evolves in time. Concerning boundary values, it seems to be physically meaningful if we assume that the fluid on the (moving) boundary has the same velocity as the wall of the vessel. Thus, we have to consider non-homogeneous Dirichlet boundary values. This non homogeneity can be handled relatively easy in the setting of strong solutions by adding an extension of the boundary data to the solution. Nevertheless, this motivates our considerations in the final Chapters 4 and 5. There we will study weak solutions to the Navier–Stokes equations in a cylindrical time-space domain with non-homogeneous boundary data.

As above, it seems to be useful to study an extension operator for the inhomogeneous boundary data β in the sense that we have to study the system

$$
\begin{aligned}
b_t - \Delta b + \nabla \tilde{p} &= 0 && \text{in } \Omega \times (0, T), \\
\operatorname{div} b &= 0, && \text{in } \Omega \times (0, T), \\
b &= \beta && \text{on } \partial\Omega \times (0, T), \\
u(0) &= 0 && \text{on } \Omega.
\end{aligned}
$$

Then, the weak solution to the Navier–Stokes equations with inhomogeneous boundary data will be constructed as a sum $u = v + b$, where b is a solution to the Stokes equations with inhomogeneous boundary data, and v is a solution to

$$
\begin{aligned}
v_t - \Delta u + (v + b) \cdot \nabla(v + b) + \nabla p &= f && \text{in } \Omega \times (0, T), \\
\operatorname{div} v &= 0 && \text{in } \Omega \times (0, T), \\
v &= 0 && \text{on } \partial\Omega \times (0, T), \\
v(0) &= u_0 && \text{on } \Omega.
\end{aligned}
$$

This system is called *perturbed Navier–Stokes system*. The crucial advantage of studying perturbed Navier–Stokes system instead of the original system are the homogeneous boundary data. The main result of Chapter 4 is the existence of a weak solution to the perturbed Navier–Stokes system in an exterior domain, which decays as

$$\|v(t)\|_{L^2(\Omega)} \leq ct^{-\frac{3}{4}+\varepsilon}$$

for any $\varepsilon > 0$, provided that the data β, u_0, and f are suitable chosen. Furthermore, the decay-rate is exponentially, if the domain is bounded. Therefore, the proof strongly depends on the construction of a weak solution and cannot directly be generalised to an arbitrary weak solution to the perturbed Navier–Stokes system.

The last Chapter 5 is dedicated to the generalisation of the decay result to a larger class of weak solutions. The main idea to prove the decay result for weak solution is to demonstrate a uniqueness result for weak solutions. In fact, we will prove that a *Serrin-type Uniqueness Theorem* holds true for the perturbed Navier–Stokes system. Then, an arbitrary weak solution which fulfils the strong energy inequality has eventually better integrability properties and this will imply, that an arbitrary weak solution which fulfils the strong energy inequality has the same decay properties as the well-chosen weak solution discussed in Chapter 4. Besides the Serrin-type uniqueness result the main ingredient of the proof will be an existence result of so-called Serrin-type strong solutions. Here, Serrin-type strong solutions are Leray-Hopf type weak solutions v such that

$$v \in L^s(0, \infty; L^q(\Omega))$$

for

$$\frac{2}{s} + \frac{3}{q} = 1.$$

This additional integrability condition is needed to ensure that the solutions can be used as a test-function in the variational formulation of weak solutions.

Acknowledgement

At this point, I would like to express my gratitude to several people who support me in writing this thesis.

First of all, let me thank my advisor Professor Dr. Reinhard Farwig for proposing the project of this thesis and his valuable support during my time as a Phd-student in Darmstadt. I am very grateful for the opportunities I had as a student of him, in particular, for my research stays in Tokyo and St. Petersburg.

I would also like to thank Professor Dr. Hideo Kozono for inviting me five times to the Waseda University and spending much time on discussions about our common research. I gained a lot from the time Professor Kozono we discussed together.

In the working group "Analysis" I would like to thank Dr. Björn Augner, Aday Celik, Thomas Eiter, Dr. Amru Hussein, Dr. Christian Komo, Professor Dr. Mads Kyed, Anke Meier-Dörnberg, Jens-Henning Möller, Dr. Martin Saal, Andreas Schmidt, and Dr. Patrick Tolksdorf for many fruitful discussions and a very nice working atmosphere in the department. In particular, let me thank Professor Kyed for refereeing my dissertation and Björn Augner, Thomas Eiter, Amru Hussein, Jens-Henning Möller, Andreas Schmidt, and Patrick Tolksdorf for their valuable proof reading.

Last but not least, I would like to thank Katharina Peier and my parents, Bettina and Robert Wegmann for supporting me in every non-mathematical problem.

Zusammenfassung

In der vorliegenden Arbeit beschäftigen wir uns mit zwei verschiedenen Problemstellung zu den Navier-Stokes Gleichungen. Die Navier-Stokes-Gleichungen

$$u_t - \nu \Delta u + u \cdot \nabla u + \nabla p = f \qquad \text{in } \Omega \times (0, T),$$
$$\operatorname{div} u = 0 \qquad \text{in } \Omega \times (0, T),$$
$$u = \beta \qquad \text{auf } \partial\Omega \times (0, T),$$
$$u(0) = u_0 \qquad \text{auf } \Omega$$

sind das heutzutage meist anerkannte System zur Modellierung der Fließgeschwindigkeit inkompressibler, Newton'scher Flüssigkeiten wie z.B. Wasser.

Betrachtet man nun ein Objekt mit sehr hoher Dichte, das in einem mit Wasser gefüllten Behälter aufgrund einer äußeren Kraft, z.B. wegen der Gravitation, sinkt, so verändert sich das mit Wasser gefüllte Gebiet. Aufgrund der hohen Dichte des sich bewegenden Objekts, ist es in erster Näherung sinnvoll, die Navier-Stokes-Gleichungen in einem nicht zylindrischen Raum-Zeit-Gebiet zu betrachten. Eine anschließende Transformation auf ein zylindrisches Raum-Zeit-Gebiet führt zu einem nicht-autonomen System, das wir im ersten Abschnitt dieser Arbeit untersuchen werden. Hierbei betrachten wir den Fall, dass zu jedem Zeitpunkt das räumliche Gebiet ein Außenraumgebiet sei.

In Kapitel 2 werden wir beweisen, dass das zugehörige linearisierte System *Maximale Regularität* in $L^s(0, \infty; L^q_\sigma)$ besitzt, falls $1 < q < \frac{3}{2}$ und $1 < s < \infty$.

Basierend auf diesem Resultat, werden wir in Kapitel 3 das nichtlineare System behandeln und zeigen, dass dieses für beliebig große Daten eine zeitlokale Lösung und für kleine Daten eine zeitglobale Lösung besitzt. Die Argumentation in diesem Kapitel basiert im Wesentlichen auf der Verwendung des Banach'schen Fixpunktsatzes und benötigt eine geeignete Abschätzung des nichtlinearem Terms.

In den beiden finalen Kapiteln dieser Arbeit werden wir uns einer anderen Fragestellung widmen. Betrachten wir nun das Navier-Stokes-System in einem zylindrischen Raum-Zeit-Gebiet mit inhomogenen Dirichlet Randdaten. Üblicherweise und so auch in dieser Arbeit, wird eine schwache Lösung u der Navier-Stokes-Gleichungen mit inhomogenen Randdaten konstruiert als $u = v + b$, wobei b eine Fortsetzung der Randdaten und v eine schwache Lösung zu einer zur Navier-Stokes-Gleichung ähnlichen Differenzialgleichung gegeben durch

$$\text{(PNST)} \quad \begin{aligned} v_t - \Delta u + (v + b) \cdot \nabla(v + b) + \nabla p &= f \quad \text{in } \Omega \times (0, T), \\ \operatorname{div} v &= 0 \quad \text{in } \Omega \times (0, T), \\ v &= 0 \quad \text{auf } \partial\Omega \times (0, T), \\ v(0) &= u_0 \quad \text{auf } \Omega \end{aligned}$$

bezeichnet. Wir werden im Folgenden zunächst beweisen, dass eine schwache Lösung v zu (PNST) existiert, die exponentiell abklingt, wenn Ω beschränkt ist, und wie $t^{-\frac{3}{4} + \varepsilon}$ für ein beliebiges $\varepsilon > 0$ unter geeigneten Annahmen an die Daten, falls Ω ein Außenraumgebiet ist. Der Beweis dieser Aussage basiert wesentlich auf dem Konstruktionsverfahren der schwachen Lösung, somit kann zunächst nur eine Aussage zur Existenz einer abklingenden Lösung getroffen werden.

Durch den Beweis einer Verallgemeinerung des Eindeutigkeitssatzes von Serrin werden wir abschließend zeigen, dass die Abklingeigenschaft einer speziell konstruierten Lösung sich unter geeigneten Voraussetzungen auf alle schwachen Lösungen, die der starken Energieungleichung genügen, übertragen lässt. Hiermit werden wir uns im letzten Kapitel beschäftigen.

CHAPTER 1

Preliminaries

Throughout this first chapter, we introduce the basic notation and concepts we will use frequently during this doctoral thesis. The first part hereof is developed to introduce the function spaces we will work with and to shortly introduce the notion of an interpolation space. As usual in the theory of parabolic equations, the solution spaces are given by Banach-space-valued Lebesgue spaces, the so-called Lebesgue–Bochner spaces and their Sobolev-type generalisation. For further reading we especially recommend the book of K. Yosida [57]. In the theory of fluid mechanics, the underlying Banach space usually consists of solenoidal vector fields as a subspace of Lebesgue functions, which have been studied for several decades; nowadays there are several textbooks dealing with them, *e.g.*, the monographs of G. Galdi [22], R. Temam [51], and H. Sohr [50].

To describe the space of initial data to parabolic differential equations, a suitable description of traces in time of the solution space is useful, and the most common way to represent those trace spaces is based on interpolation theory. We will shortly introduce the corresponding notation of H. Triebel [53] and J. Bergh and J. Löfström [3] and apply it in Section 1.3.

The well -posedness of a parabolic differential equation is strongly connected to the property of the differential operator to be generator of a C^0-semigroup, and the regularity properties correlate to the behaviour of the resolvent of the differential operator. An easy to read introduction into the theory of semigroups can be found in the book of K. Engel and R. Nagel [12]. Regularity properties of the resolvent will also yield a possibility to introduce a functional calculus. Another functional calculus will yield a decay result for solutions to linear equations, provided that the generator is an injective, self-adjoint operator, see Subsection 1.2.1. This tool will be necessary to prove the decay results in the final Chapters 4 and 5.

One of the best regularity properties of parabolic differential operators is that the operator is a homeomorphism of the solution space to the space of right-hand sides and initial values. This is usually called maximal regularity. The main breakthrough in this theory was developed by L. Weis [55,56], when he proved that an operator has maximal regularity if and only if it is $\mathcal{R}$-sectorial. Together with some perturbation results concerning maximal regularity, this will be described in Section 1.3.

Strongly connected to the question whether a differential operator A has maximal regularity or not is to find sufficient conditions on A such that $\partial_t + A$ is a closed operator. Regularity theory to $\partial_t + A$ also leads to estimates for $\partial_t^\alpha A^{1-\alpha}$, which is essential to obtain existence results for quasi-linear differential equations.

Finally, since all of this dissertation deals with the Stokes and Navier–Stokes equations in bounded and especially in exterior domains, we have to introduce the corresponding operators and, of course, the main estimates we will use in the subsequent chapters.

1.1 Function Spaces, Interpolation Theory and Basic Notation

In this first section we introduce important function spaces and some basic notation we will use frequently in this dissertation. We will denote by $L^p(\Omega; X)$ for some domain $\Omega \subset \mathbb{R}^n$, $n \in \mathbb{N}$, and some real or complex Banach space X the *Lebesgue–Bochner space* of all measurable functions $f \colon \Omega \to X$ such that the real-valued function $x \mapsto \|f(x)\|_X$ is an element of the Lebesgue space $L^p(\Omega; \mathbb{R})$. The corresponding norm will be denoted by $\| \cdot \|_{L^p(\Omega;X)}$, and if there is no possible confusion we will write $\|f\|_p$. The most important Lebesgue–Bochner spaces will be those where X is a real-valued Lebesgue space and the functions are defined on an interval I.

The norm to those $L^s(I; L^q(\Omega))$ spaces will be denoted by

$$\|f\|_{L^s(I;L^q(\Omega))} = \|f\|_{q,s,I} = \|f\|_{q,s}.$$

If $I = [0, T)$, we may replace I by T in the short notation of the norm.

As usual, the *Sobolev space* of k-th order, $k \in \mathbb{N}_0$, will be written as $W^{k,p}(\Omega)$ or $W^{k,p}(\Omega; X)$. Since the main part of this dissertation deals with unbounded domains, we also have to introduce the *homogeneous spaces* $\dot{W}^{k,p}(\Omega)$. This space consists of all measurable functions, which have distributional derivatives of k-th order that are represented by $L^p(\Omega)$ functions. This space is endowed with the semi-norm

$$\|u\|_{\dot{W}^{k,p}(\Omega)} := \sum_{|\alpha|=k} \|\partial^\alpha u\|_p.$$

Here α denotes a multi-index $\alpha \in \mathbb{N}_0^n$ such that $|\alpha| = \sum_{j=1}^n \alpha_j = k$.

The trace space of $W^{k,q}(\Omega)$ is denoted by $W^{k-\frac{1}{q},q}(\partial\Omega)$ provided that $\partial\Omega$ is sufficiently smooth.

It is well-known that the set of all smooth compactly supported functions $C_0^\infty(\Omega)$ is dense in $L^p(\Omega)$, $1 \le p < \infty$. The closure of this space in $W^{k,p}(\Omega)$ will be denoted by $W_0^{k,p}(\Omega)$. Furthermore, let

$$C_{0,\sigma}^\infty(\Omega) := \{\varphi \in C_0^\infty(\Omega) \mid \operatorname{div}\varphi = 0\}.$$

The closure of $C_{0,\sigma}^\infty(\Omega)$ with respect to the norm $\|\cdot\|_p$ defines a closed subspace of $L^p(\Omega)$ and will be denoted by $L_\sigma^p(\Omega)$.

Let Ω be a domain with $\partial\Omega \in C^{1,1}$, and $\varphi\colon \Omega \to \mathbb{R}$ and $\psi\colon \Omega \to \mathbb{R}^n$ be smooth functions. Then the Divergence Theorem implies

$$\langle \varphi, n \cdot \psi \rangle_{\partial\Omega} := \int_{\partial\Omega} \varphi(n \cdot \psi)\, d\sigma = \int_\Omega (\nabla\varphi) \cdot \psi\, dx + \int_\Omega \varphi \operatorname{div}\psi\, dx$$
$$= \langle \nabla\varphi, \psi \rangle_\Omega + \langle \varphi, \operatorname{div}\psi \rangle_\Omega,$$

where n denotes the outer normal vector. Now, for all $\varphi \in W^{1,q'}(\Omega)$ the right-hand side is well-defined for all $\psi \in L^q(\Omega)$ such that $\operatorname{div}\psi \in L^r(\Omega)$ and $\frac{1}{r} = \frac{1}{q} + \frac{1}{n}$. In this case, the right-hand defines a functional in $W^{1,q'}(\Omega)$ due to Sobolev embedding. Hence we can define for functions ψ as above a *normal trace* and that trace space will be denoted by $W^{-\frac{1}{q},q}(\partial\Omega)$. In particular, there is a normal trace for functions in $L_\sigma^q(\Omega)$.

It is well-known, that if $\partial\Omega$ is sufficiently smooth, then

$$(1.1) \qquad L_\sigma^p(\Omega) = \{u \in L^p(\Omega) \mid w\text{-div}\, u = 0,\ u \cdot n = 0 \text{ in } W^{-\frac{1}{q},q}(\partial\Omega)\}.$$

Here, w-div denotes the *weak divergence* and it is defined as

$$w\text{-div}\colon L^p(\Omega) \to W^{-1,p}(\Omega), u \mapsto \left(\varphi \mapsto -\langle u, \nabla\varphi\rangle\right).$$

Let us argue shortly that the inclusion "$\subset$" in (1.1) provided that $p > n$ is obvious. Let $u \in L_\sigma^p(\Omega)$. Then there exists a sequence $(u_n)_n \subset C_{0,\sigma}^\infty(\Omega)$ such that $u_n \to u$ in $L^p(\Omega)$. Thus we have

$$0 = w\text{-div}\, u_n \to w\text{-div}\, u$$

and we see that the distributional divergence of u is in $L^r(\Omega)$, $\frac{1}{r} = -\frac{1}{n} + \frac{1}{p}$. Finally, the trace result above implies $0 = u \cdot n \in W^{-\frac{1}{p},p}(\partial\Omega)$. For the inclusion "$\supset$" we refer to [22, 50].

Let us start to introduce the notion of complex and real interpolation. A couple of Banach spaces (X, Y) is called *interpolation couple*, if there exists a topological Hausdorff space $\mathcal{V}$ such that $X \cup Y \subset \mathcal{V}$. Hence $X + Y := \{x + y \mid x \in X, y \in Y\} \subset \mathcal{V}$ is well-defined. We endow the space $X \cap Y$ with the norm

$$\|z\|_{X \cap Y} := \max\{\|z\|_X, \|z\|_Y\}$$

and the space $X + Y$ with

$$\|z\|_{X+Y} := \inf\{\|x\|_X + \|y\|_Y \mid x + y = z, x \in X, y \in Y\}.$$

To define real interpolation spaces, let us introduce the so-called K-*functional*.

Definition 1.1.1. Let X, Y denote an interpolation couple. Let us define the K-functional by

$$K(\cdot, \cdot, X, Y)\colon [0, \infty) \times (X + Y) \to [0, \infty),$$
$$(t, z) \mapsto K(t, z, X, Y) := \inf\{\|x\|_X + t\|y\|_Y \mid z = x + y, x \in X, y \in Y\}.$$

For simplicity we write $K(t, z)$ instead of $K(t, z, X, Y)$ if there is no reason for confusion.

Furthermore, let us denote by $L^p_\star(0, \infty)$, $1 \leq p < \infty$, the space of those measurable functions f such that

$$\|f\|_{p,\star} := \left(\int_0^\infty |f(t)|^p \frac{\mathrm{d}t}{t} \right)^{\frac{1}{p}} < \infty,$$

and $L^\infty_\star(0, \infty) := L^\infty(0, \infty)$. Let X be a Banach space and let us also define the space $L^p_\star(0, \infty; X)$ of all measurable functions $f \colon (0, \infty) \to X$ such that $\|f(\cdot)\|_X \in L^p_\star(0, \infty)$ endowed with the canonical norm. For reasons we will see later, let us mention the following result, which can be found in [37, Corollary A.13].

Lemma 1.1.2. *Let X be a Banach space and $u \colon (0, \infty) \to X$ be a measurable function such that $t \mapsto u_\theta(t) := t^\theta u(t)$ belongs to $L^p_\star(0, \infty; X)$ for some $\theta \in (0, 1)$. Then the mean value*

$$v(t) := \frac{1}{t} \int_0^t u(\tau) \, \mathrm{d}\tau$$

defines a function such that $t \mapsto v_\theta(t) := t^\theta v(t)$ is an element of $L^p_\star(0, \infty; X)$ and the estimate

$$\|v_\theta\|_{L^p_\star(0,\infty;X)} \leq \frac{1}{1-\theta} \|u_\theta\|_{L^p_\star(0,\infty;X)}$$

holds true.

Now let us define real interpolation spaces.

Definition 1.1.3. Let X, Y be a interpolation couple, $1 \leq p \leq \infty$, and $\theta \in (0, 1)$. Then we define the *real interpolation space of X and Y with parameter θ and p* by

$$(X, Y)_{\theta, p} := \{ z \in X + Y \mid t \mapsto t^{-\theta} K(t, z) \in L^p_\star(0, \infty) \}$$

endowed with the norm

$$\|z\|_{(X,Y)_{\theta,p}} := \|t \mapsto t^{-\theta} K(t, z)\|_{L^p_\star(0,\infty)}.$$

Real interpolation will be important to characterise the space of initial values later on. However, we will also make use of the complex interpolation method. For this concept we introduce some notation first. Let us define the strip

$$S := \{z \in \mathbb{C} \mid \operatorname{Re}(z) \in (0,1)\},$$

and let X, Y denote an interpolation couple. Let $\mathcal{F}(X,Y)$ denote the set of all analytic functions $f \colon S \to X + Y$ such that there exists a continuous extension of f on $\overline{S}$ such that $f(i\mathbb{R}) \subset X$ and $f(1 + i\mathbb{R}) \subset Y$. We endow this space with the norm

$$\|f\|_{\mathcal{F}(X,Y)} := \sup_{t\in\mathbb{R}}(\|f(it)\|_X + \|f(1+it)\|_Y).$$

Here, $\|f\|_{\mathcal{F}(X,Y)}$ defines a norm due to hadamard's Three-Lines Theorem.

Definition 1.1.4. Let X, Y be an interpolation couple and $\theta \in (0,1)$. Then the *complex interpolation space of parameter θ* $(X,Y)_\theta$ is defined as

$$(X,Y)_\theta := \{f(\theta) \mid f \in \mathcal{F}(X,Y)\}$$

endowed with the norm

$$\|z\|_{(X,Y)_\theta} := \inf\{\|f\|_{\mathcal{F}(X,Y)} \mid f(\theta) = z\}.$$

The theory of complex interpolation will be important to characterise the domains of fractional powers of a large class of linear operators.

1.2 A Short Glimpse on Operator Theory

Let us start to present the most important notions from operator theory we will use frequently throughout this dissertation. Let X be a Banach space and let A denote a closed operator $A \colon \mathcal{D}(A) \subset X \to X$. Whenever it might be confusing not to mention the domain explicitly, we will denote the operator by $(A, \mathcal{D}(A))$. The *resolvent set* ρ_A of a closed operator A is defined by

$$\rho_A := \{\lambda \in \mathbb{C} \mid (\lambda - A)^{-1} \in \mathcal{L}(X)\}.$$

Here and hereafter $\mathcal{L}(X)$ denotes the set of linear and continuous operators from X to X.

The complement $\sigma_A := \mathbb{C} \setminus \rho_A$ is called the *spectrum* of A. To define one of the most important classes of closed operators, let us introduce the sector

$$S_\theta := \{\lambda \in \mathbb{C} \setminus \{0\} \mid |\arg(\lambda)| < \theta\}, \ 0 < \theta \leq \pi,$$

and

$$S_{\theta,\delta} := S_\theta \setminus B_\delta(0).$$

Definition 1.2.1. A closed, injective linear operator A on a Banach space X is called *sectorial operator* if A is densely defined, has dense range, and if $\sigma_A \subset \overline{S_\omega}$ for some $\omega \in (0,\pi)$ and for every $\theta \in (\omega,\pi)$ there exists a constant c_θ such that

$$(1.2) \qquad\qquad \|\lambda(\lambda + A)^{-1}\|_{\mathcal{L}(X)} \leq c_\theta$$

holds true for all $\lambda \in S_{\pi-\theta}$. The infimum of those ω is called *spectral angle* and will be denoted by ϕ_A.

It is known that if an injective closed operator A fulfils (1.2) on a sector S_θ, then A is densely defined and has dense range provided that the Banach space X is reflexive, see [30, Proposition 2.1.1]. Furthermore, we obtain that $\mathcal{D}(A^k) \cap \text{Im}(A^k)$ is dense for all $k \in \mathbb{N}$.

In the next part, we would like to briefly describe how one can define a functional calculus for sectorial operators. Let $\mathcal{H}(S_\theta)$ denote the set of holomorphic functions $f \colon S_\theta \to \mathbb{C}$. Let A denote a sectorial operator and let $\theta > \phi_A$. If A is a bounded operator with $0 \in \rho(A)$, and $f \in \mathcal{H}(S_\theta)$ with $\theta > \phi_A$ one can define $f(A)$ by

$$(1.3) \qquad\qquad f(A) := \frac{1}{2\pi i} \int_\Gamma f(\lambda)(\lambda - A)^{-1} \mathrm{d}\lambda.$$

Here Γ denotes a path which surrounds the spectrum of A counter-clockwise and lies in S_θ and Γ is chosen such that the length of the path $\Gamma \cap B_R$ can be estimated by R for large R. Due to Cauchy's Theorem, $f(A)$ does not depend on Γ.

Let us define

$$\mathcal{H}_0(S_\theta) := \bigcup_{\alpha,\beta<0} \mathcal{H}_{\alpha,\beta}(S_\theta),$$
$$\mathcal{H}_{\alpha,\beta}(S_\theta) := \{f \in \mathcal{H}(S_\theta) \mid \|f\|_{\alpha,\beta} < \infty\},$$

where the semi-norm above is defined as

$$\|f\|_{\alpha,\beta} := \sup\{|\lambda^\alpha f(\lambda)| + |\lambda^{-\beta} f(\lambda)| \mid \lambda \in S_\theta\}.$$

The estimate (1.2) for the resolvent of a sectorial operator and the definition of $\mathcal{H}_0(S_\theta)$ imply that the integral in (1.3) exists for all invertible, sectorial operators A and functions $f \in \mathcal{H}_0(S_\theta)$, $\theta > \phi_A$. Furthermore, if $0 \notin \rho_A$ we can consider the approximations

$$A_\varepsilon := (\varepsilon + A)(1 + \varepsilon A)^{-1}, \ \varepsilon > 0,$$

and one can prove easily that A_ε is a bounded, invertible operator and converges strongly to A for all $x \in \mathcal{D}(A)$. Furthermore, A_ε is a sectorial operator on a larger sector than A, and for all $\lambda \in S_{\pi-\phi_A}$ we have

$$\lim_{\varepsilon \to 0} \|(\lambda + A_\varepsilon)^{-1} - (\lambda + A)^{-1}\|_{\mathcal{L}(X)} = 0.$$

It is well-known that for all $f \in \mathcal{H}_0(S_\theta)$ and all sectorial operators A the operator $f(A_\varepsilon)$ converge as $\varepsilon \to 0$ to some operator in $\mathcal{L}(X)$ with respect to the topology of $\mathcal{L}(X)$, and hence we can define $f(A)$ as this limit. Finally, we would like to enlarge the class of functions f such that we can define $f(A)$. Let ψ denote the holomorphic function $\psi(\lambda) := \lambda(1 + \lambda)^{-2}$. Then $\psi(A) = A(1 + A)^{-2}$ and $(\psi(A))^{-1} = 2 + A + A^{-1}$ are well-defined on $\mathcal{D}(A) \cap \text{Im}(A)$. Hence let $f \in \mathcal{H}(S_\theta)$ be a holomorphic function such that $\psi^k f \in \mathcal{H}_0(S_\theta)$ for some $k \in \mathbb{N}$; then we define

(1.4)
$$f(A) := \psi(A)^{-k}(\psi^k f)(A),$$
$$\mathcal{D}(f(A)) := \{x \in X \mid (\psi^k f)(A)x \in \mathcal{D}(A^k) \cap \text{Im}(A^k)\}.$$

We obtain the following Proposition [10, Theorem 2.1].

Proposition 1.2.2. *Let X be a Banach space and A be a sectorial operator. Then the functional calculus defined by* (1.4) *is well-defined for all*

$$f \in \bigcup_{\alpha > 0} \mathcal{H}_{\alpha,\alpha}(S_\theta),$$

$\theta > \phi_A$, *and defines a closed linear operator* $f(A)$.

In particular, Proposition 1.2.2 enables us to define fractional powers A^α of a sectorial operator A. Furthermore, the function $\psi\colon z \mapsto z^\alpha$ for some $\alpha \in \mathbb{C}$ defines an analytic function on the sliced complex plane, and due to Proposition 1.2.2 we can define $\psi(A)$ for any sectorial operator A. Especially, we obtain that $A^{i\alpha}$ is well defined for all $\alpha \in \mathbb{R}$ and defines a family of closed operators with properties like a group of operators.

Definition 1.2.3. A sectorial operator A is said to admit *bounded imaginary powers* if $A^{i\alpha} \in \mathcal{L}(X)$ for all $\alpha \in \mathbb{R}$ and there exists a $c > 0$ such that

$$\|A^{i\alpha}\|_{\mathcal{L}(X)} \leq c \text{ for all } |\alpha| \leq 1.$$

The *power angle* θ_A is defined by

$$\theta_A := \limsup_{|\alpha| \to \infty} \frac{1}{|\alpha|} \log(\|A^{i\alpha}\|_{\mathcal{L}(X)}).$$

The class of all sectorial operators with bounded imaginary powers will be denoted by $\mathcal{BIP}$.

Operators with bounded imaginary powers have several very important properties. Throughout this dissertation we will need especially two of them. One is connected to prove a sufficient condition whenever the sum of closed operators defines a closed operator and will presented later on. Furthermore, the property of $\mathcal{BIP}$ is helpful to characterise $\mathcal{D}(A^\alpha)$, see [54, Theorem 1.15.3].

Proposition 1.2.4. *Let X be a Banach space and $A \in \mathcal{BIP}$. Then*

$$[\mathcal{D}(A^{\alpha_1}), \mathcal{D}(A^{\alpha_2})]_\theta = \mathcal{D}(A^\alpha)$$

for all $\alpha_1, \alpha_2 \in \mathbb{C}$ such that $0 \leq \operatorname{Re}(\alpha_1) < \operatorname{Re}(\alpha_2) < \infty$, $\theta \in (0,1)$, and $\alpha = \theta\alpha_1 + (1-\theta)\alpha_2$.

Definition 1.2.5. A sectorial operator A with spectral angle ϕ_A is said to admit a *bounded $\mathcal{H}^\infty$-calculus* (on the sector S_θ), if there exists $c > 0$ and $\theta \in (\phi_A, \pi]$ such that

$$(1.5) \qquad \|f(A)\|_{\mathcal{L}(X)} \leq c\|f\|_{\mathrm{L}^\infty(S_\theta)} \text{ for all } f \in \mathcal{H}_0^\infty(S_\theta).$$

The smallest angle θ such that there exists a constant $c = c(\theta)$ such that (1.5) holds is called $\mathcal{H}^\infty$-*angle* and will be denoted by ϕ_A^∞. Furthermore, the class of all sectorial operators with bounded $\mathcal{H}^\infty$-calculus will be denotes by $\mathcal{H}^\infty$.

It is well-known that $\mathcal{H}^\infty \subset \mathcal{BIP}$ and

$$\phi_A^\infty \geq \theta_A \geq \phi_A.$$

1.2.1 The Hilbert Space Case

There are several approaches to define a functional calculus, *i.e.*, to define $f(A)$ for a class of operators A and a class of functions f. In this section, so far, we described an approach to define $f(A)$ for sectorial operators A and analytic functions f. In this subsection we would like to introduce a functional calculus for self-adjoint operators since this will be crucial to prove the decay results in Chapter 4 and Chapter 5.

Let H denote a *Hilbert space* and A a closed *positive self-adjoint* operator on H. Then there exists a family of orthogonal projections $(E_\lambda)_{\lambda \geq 0}$ with the following properties:

(i) $E_\lambda E_\mu = E_\mu E_\lambda = E_\lambda$ for all $0 \leq \lambda \leq \mu < \infty$.

(ii) $E_\lambda x = \lim_{\mu \to \lambda, \mu < \lambda} E_\mu x$ for all $x \in H$.

(iii) $E_0 = 0$ and $\lim_{\lambda \to \infty} E_\lambda x = x$ for all $x \in H$.

A family of orthogonal projections with those properties is called *resolution of the identity*.

Using a Riemann–Stieltjes integral, we define

$$\int_0^b g(\lambda) \, \mathrm{d}\|E_\lambda v\|^2, \ v \in H, \ 0 < b < \infty,$$

as a limit of Riemann–Stieltjes sums

$$\sum_{j=1}^{m} g(\lambda_j)\big(\|E_{\lambda_j}v\|^2 - \|E_{\lambda_{j-1}}v\|^2\big) = \sum_{j=1}^{m} g(\lambda_j)\|E_{\lambda_j}v - E_{\lambda_{j-1}}v\|^2$$

with a partition

$$0 = \lambda_0 < \lambda_1 < \ldots < \lambda_m = b,$$

such that its width length tends to 0. Of course, this can be done for a large class of functions including all continuous functions $g \in C^0([0,\infty);\mathbb{R})$. In that case, also

$$\int_0^b g(\lambda)\mathrm{d}E_\lambda v, \ v \in H,$$

is well-defined as a limit of Riemann sums, and we have

$$\left\| \int_0^b g(\lambda)\,\mathrm{d}E_\lambda v \right\|^2 = \int_0^b g^2(\lambda)\mathrm{d}\|E_\lambda v\|^2.$$

Furthermore, we define

$$\int_0^\infty g(\lambda)\mathrm{d}E_\lambda$$

as the improper integral

$$\lim_{b\to\infty} \int_0^b g(\lambda)\mathrm{d}E_\lambda$$

and this limit exists w.r.t. the strong operator topology. It is well-known that for a closed self-adjoint operator A there exists a resolution of the identity E_A to A, the so-called *spectral representation*, such that

(1.6) $$A = \int_0^\infty \lambda \mathrm{d}E_A(\lambda)$$

and

$$\mathcal{D}(A) = \Big\{ x \in H \mid \int_0^\infty |\lambda|^2 \mathrm{d}\|E_A(\lambda)x\|^2 < \infty \Big\}.$$

Since for every $x \in H$ the function $\lambda \mapsto \|E_A(\lambda)x\|$ is by assumption (i) monotonically increasing, we see that $Ax = 0$ if and only if for all $\varepsilon > 0$ there exists λ_0 such that

$$\|E_A(\lambda)x\| < \varepsilon \text{ for all } \lambda \leq \lambda_0,$$

and hence we obtain that the statements

(i) A is injective,

(ii) The map $\lambda \mapsto E_A(\lambda)$ is strongly continuous at $\lambda = 0$,

are equivalent.

Of course, the functional calculus described above enables us to define the semigroup generated by a self-adjoint operator by

$$e^{-tA} := \int_0^\infty e^{-t\lambda} \mathrm{d}E_A(\lambda).$$

We obtain the following result.

Lemma 1.2.6. *Let A be an injective self-adjoint operator on the Banach space H. Then*

$$\lim_{t \to \infty} e^{-tA} x = 0$$

for all $x \in H$.

Proof. Since A is injective, we have that $\lambda \mapsto E_A(\lambda)$ is continuous at 0. Furthermore, $e^{-t\lambda}$ tends to 0 for all $\lambda > 0$ as $t \to \infty$. Thus, the theorem by Lebesgue on dominated convergence yields the result. $\square$

Note that the special case in which A is the Stokes operator on a domain $\Omega \subset \mathbb{R}^3$ was considered by Masuda [38].

1.3 Maximal Regularity

In this section we introduce the notion of *Maximal Regularity* . Throughout this section, let X denote a Banach space and let $(A, \mathcal{D}(A))$ denote a linear operator on X such that $-A$ generates an analytic semigroup. Furthermore, let $I \subset [0, \infty)$ denote an interval such that $0 \in I$. For a precise definition, we consider the *abstract Cauchy problem*

(1.7)
$$\partial_t u + Au = f \text{ in } I,$$
$$u(0) = u_0$$

with $f \in \mathrm{L}^p(I; X)$, $1 < p < \infty$. Let us first restrict to the case $u_0 = 0$. Since $-A$ generates an analytic semigroup, we know that there exists a unique mild solution u to (1.7), which is given by

$$u(t) = \int_0^t e^{-(t-s)A} f(s) \mathrm{d}s.$$

Definition 1.3.1. Let X be a Banach space, let $1 < p < \infty$ and let $I \subset [0, \infty)$ with $0 \in I$ denote an interval. Furthermore, let $-A$ be a generator of an analytic semigroup. We say that the operator A has *maximal* L^p-*regularity*, if for all $f \in L^p(I; X)$ the unique mild solution u to (1.7) is almost everywhere differentiable and there exists $c > 0$ such that

$$(1.8) \qquad \|\partial_t u\|_{L^p(I;X)} + \|Au\|_{L^p(I;X)} \leq c \|f\|_{L^p(I;X)}$$

holds. The set of all operators with maximal L^p-regularity on the Banach space X in the time interval I will be denoted by $\mathcal{MR}_p(I, X)$. Moreover, the set $\mathcal{MR}_p(I, X, c)$ denotes the set of all $A \in \mathcal{MR}_p(I, X)$ such that the constant in estimate (1.8) does not exceed c. If $I = [0, \infty)$, we will simply write $\mathcal{MR}_p(X)$ or $\mathcal{MR}_p(X, c)$.

Note that we do not require in the definition of maximal regularity that we have control of $\|u\|_{L^p(I;X)}$. But it is easy to see that this is automatically fulfilled, if I is finite or A is invertible.

Lemma 1.3.2. *Let X be a Banach space, let $1 < p < \infty$, and let $I \subset [0, \infty)$ with $0 \in I$ denote an interval. Furthermore, let $-A$ be a generator of a bounded analytic semigroup. Furthermore, assume that A is invertible or I is finite. Then the following statements are equivalent:*

(i) *The operator A has maximal L^p-regularity.*

(ii) *The differential operator*

$$(\partial_t + A) \colon W^{1,p}(I; X) \cap L^p(I; \mathcal{D}(A)) \to L^p(I; X)$$

is a homeomorphism.

Proof. Assume that A has maximal L^p-regularity. If A is invertible, then the boundedness of A^{-1} and $Au \in L^p(I; X)$ implies that $u \in L^p(I; X)$. If the time interval I is finite, we consider the operator $\tilde{A} := A + \mu$ for some $\mu > 0$ and obtain that u solves (1.7) if and only if $e^{-\mu t}u(t)$ solves (1.7) with the operator $\tilde{A}$ and $\tilde{f} := e^{-\mu t}f$. Since $\tilde{A}$ is invertible and $\tilde{A}$ has obviously maximal L^p-regularity since A has maximal L^p-regularity, we obtain the statement.

Note that the space $W^{1,p}(I;X) \cap L^p(I;\mathcal{D}(A))$ is a Banach space. Hence if the differential operator $(\partial_t + A)$ is a homeomorphism, we obtain that its inverse is continuous due to the Closed Graph Theorem. Hence (ii) implies (i). □

Next, let us extend the notion of maximal L^p-regularity to the case of non-zero initial values. The canonical space of initial values is, of course, the space of all traces of functions in the solution space, *i.e.*, $u(0)$ for all u such that $\partial_t u, Au \in L^p(I;X)$. We will prove that this space coincides with equivalent norms with a real interpolation space. The proof is due to H. Butzer and P. Berens [9] if we endow the space $\mathcal{D}(A)$ with the graph norm $\|a\|_X + \|Aa\|_X$. We will prove that their result extends to the case if we consider the closure of $\mathcal{D}(A)$ with respect to the homogeneous norm $\|Aa\|_X$. Note that it is remarked in several papers that the proof of Butzer and Berens extends to the case, where $\mathcal{D}(A)$ is endowed with the homogeneous norm, see for instance [26, Remark 2.4].

Let A denote an injective operator and let us introduce the space $\mathscr{I}^p_{A,T}$ as the space of all $a \in X$ such that

$$(1.9) \qquad \|a\|_{\mathscr{I}^p_{A,T}} := \left(\int_0^T \|Ae^{-tA}a\|^p_X \mathrm{d}t \right)^{\frac{1}{p}} < \infty.$$

Furthermore, let $W^p_{A,T}$ denote the solution space to (1.7), *i.e.*, we have

$$W^p_{A,T} := \Big\{ u\colon [0,T] \to X \mid u \text{ is measurable, } u(t) \in \mathcal{D}(A)$$
$$\text{for almost all } t \in (0,T) \text{ and } \partial_t u, Au \in L^p(0,T;X) \Big\}$$

endowed with the canonical norm

$$(1.10) \qquad \|u\|_{W^p_{A,T}} := \|\partial_t u\|_{L^p(0,T;X)} + \|Au\|_{L^p(0,T;X)}.$$

Then we may define the trace space $\mathscr{F}^p_{A,T}$ by

$$(1.11) \quad \mathscr{F}^p_{A,T} := \{u(0) \mid u \in W^p_{A,T}\}, \ \|a\|_{\mathscr{F}^p_{A,T}} := \inf\{\|u\|_{W^p_{A,T}} \mid u(0) = a\}.$$

If there is no reason for confusion, we will omit the index T and write W^p_A, $\mathscr{F}^p_A$, and $\mathscr{I}^p_A$.

Lemma 1.3.3. *Let X be a Banach space, let $1 < p < \infty$ and let $I = [0, \infty)$. Furthermore, let $-A$ be injective and a generator of a bounded analytic semigroup and let A have maximal regularity. Then the spaces $\mathscr{I}_A^p$ and $\mathscr{F}_A^p$ coincide with the real interpolation space $(X, \widehat{\mathcal{D}(A)})_{1-\frac{1}{p}, p}$ with equivalent norms.*

Here, $\widehat{\mathcal{D}(A)}$ denotes the closure of the space $\mathcal{D}(A)$ w.r.t. the homogeneous norm.

As discussed, the proof of the equivalence of $\mathscr{I}_A^p$ and the real interpolation space was first proven by Butzer and Berens in [9] if the domain of A is endowed with the graph norm. We will see that their arguments can be adopted to the homogeneous case. The proof of the equivalence of $\mathscr{I}_A^p$ and $\mathscr{F}_A^p$ was carried out in [23] and will be presented here for the sake of completeness.

Proof. Let us first assume that $x \in X$ such that $\|x\|_{\mathscr{I}_A^p}$ is finite. Thus we have $t \mapsto \|Ae^{-tA}x\|_X \in \mathrm{L}^p(0, \infty)$ or equivalently

$$t \mapsto \|t^{\frac{1}{p}} Ae^{-tA}x\|_X \in \mathrm{L}_*^p(0, \infty).$$

Then by Lemma 1.1.2 we obtain that

$$t \mapsto \varphi(t) := t^{\frac{1}{p}-1} \int_0^t \|Ae^{-sA}x\| \mathrm{d}s \in \mathrm{L}_*^p(0, \infty)$$

and that $\|\varphi\|_{\mathrm{L}_*^p(0,\infty)} \leq c \|Ae^{-\cdot A}x\|_{\mathrm{L}^p(0,\infty;X)}$. Since

$$x = x - e^{-tA}x + e^{-tA}x = -\int_0^t Ae^{-sA}x \, \mathrm{d}s + e^{tA}x,$$

and the first addend is an element in X and the second in $\mathcal{D}(A)$, we obtain

$$t^{\frac{1}{p}-1} K(t, x, X, \widehat{\mathcal{D}(A)}) \leq t^{\frac{1}{p}-1} \int_0^t \|Ae^{sA}x\|_X \mathrm{d}s + t^{\frac{1}{p}} \|Ae^{tA}x\|_X.$$

Since both terms are in $\mathrm{L}_*^p(0, \infty)$, we obtain $\mathscr{I}_A^p \subset (X, \widehat{\mathcal{D}(A)})_{1-\frac{1}{p}, p}$ with a continuous embedding.

If we assume that $x \in (X, \widehat{\mathcal{D}(A)})_{1-\frac{1}{p}, p}$, $a \in X$, and $b \in \widehat{\mathcal{D}(A)}$ with $x = a + b$, we see that

$$\|Ae^{-tA}x\|_X \leq \|Ae^{-tA}a\|_X + \|e^{-tA}Ab\|_X \leq c(t^{-1}\|a\|_X + \|Ab\|_X).$$

Thus, we have

$$\|Ae^{-tA}x\|_X \leq ct^{-1}K(t, x, X, \widehat{\mathcal{D}(A)})$$

and, since the right-hand side is in $L^p(0, \infty)$, we get that

$$(X, \widehat{\mathcal{D}(A)})_{1-\frac{1}{p}, p} \subset \mathscr{I}_A^p$$

and that the embedding is also continuous.

Let $x \in \mathscr{I}_A^p$ be given. Then the function $u(t) := e^{-tA}x$ is in W_A^p and $\|u\|_{W_A^p} \leq 2\|x\|_{\mathscr{I}_A^p}$. Hence we have $\|x\|_{\mathscr{F}_A^p} \leq 2\|x\|_{\mathscr{I}_A^p}$.

Let vice versa $x \in \mathscr{F}_A^p$. Then there exists $u \in W_A^p$ such that $u(0) = x$ and $\|x\|_{\mathscr{F}_A^p} \leq 2\|u\|_{W_A^p}$. We write

$$u(t) = e^{-tA}x + \int_0^t e^{-(t-s)A}(\partial_t u(s) + Au(s))\mathrm{d}s =: u_1(t) + u_2(t).$$

Thus, u_2 solves the system

$$\partial_t u_2 + Au_2 = \partial_t u + Au,$$
$$u_2(0) = 0.$$

Since the operator A has maximal regularity, we obtain that $u_2 \in W_A^p$ and in addition $u_1 = u - u_2 \in W_A^p$. Furthermore, we have

$$\|Au_2\|_{L^p(I;X)} \leq c\Big(\|\partial_t u\|_{L^p(I;X)} + \|Au\|_{L^p(I;X)}\Big).$$

Hence we obtain

$$(1.12) \quad \|x\|_{\mathscr{I}_A^p} = \|Au_1\|_{L^p(I;X)} \leq \|Au\|_{L^p(I;X)} + \|Au_2\|_{L^p(I;X)} \leq c\|u\|_{W_A^p}.$$

The proof is complete. $\qquad\qquad\qquad\qquad\qquad\qquad\qquad\qquad\qquad\qquad \square$

Remark 1.3.4. (i) Note that the proof of the equivalence of the norms of $\mathscr{I}_{A,T}^p$ and $\mathscr{F}_{A,T}^p$ remains true if we consider finite time intervals $[0, T)$, $0 < T < \infty$. The corresponding proof carries over one-to-one from the case $T = \infty$.

(ii) Let us note that we proved for an initial value x the estimates

$$\|x\|_{\mathscr{F}_A^p} \leq 2\|x\|_{\mathscr{I}_A^p},$$

hence the constant in this estimate obviously does not depend on A. Also in the reverse estimate we obtain using (1.12) that the constant depends only on the constant in the maximal L^p-regularity estimate (1.8) of A (with initial value 0).

The last lemma enables us to extend the notion of maximal regularity to the case of a non-zero initial value.

Lemma 1.3.5. *Let X be a Banach space, let $1 < p < \infty$, and let $I = [0, \infty)$. Furthermore, let A be an operator with maximal regularity. Then for all $f \in L^p(0, \infty; X)$ and all $u_0 \in (X, \widehat{\mathcal{D}(A)})_{1-\frac{1}{p}, p}$ there exists a solution u to (1.7) such that*

$$\|\partial_t u\|_{L^p(0,\infty;X)} + \|Au\|_{L^p(0,\infty;X)} \leq c\big(\|f\|_{L^p(0,\infty;X)} + \|u_0\|_{(X, \widehat{\mathcal{D}(A)})_{1-\frac{1}{p}, p}}\big)$$

holds true.

Proof. Due to Lemma 1.3.3 there exists a function $\tilde{u} \in W_A^p$ such that $\tilde{u}(0) = u_0$ and $\|u_0\|_{\mathscr{F}_A^p} \leq 2\|u\|_{W_A^p}$. Then, since A has maximal regularity, there exists a solution to

$$\partial_t \bar{u} + A\bar{u} = f - \partial_t \tilde{u} - A\tilde{u},$$
$$\bar{u}(0) = 0.$$

Let us define $u := \bar{u} + \tilde{u}$. Then obviously u is a solution to (1.7) and we obtain

$$\|\partial_t u\|_{L^p(0,\infty;X)} + \|Au\|_{L^p(0,\infty;X)}$$
$$\leq \|\partial_t \bar{u}\|_{L^p(0,\infty;X)} + \|A\bar{u}\|_{L^p(0,\infty;X)}$$
$$+ \|\partial_t \tilde{u}\|_{L^p(0,\infty;X)} + \|A\tilde{u}\|_{L^p(0,\infty;X)}$$
$$\leq c\big(\|f\|_{L^p(0,\infty;X)} + \|\partial_t \tilde{u}\|_{L^p(0,\infty;X)} + \|A\tilde{u}\|_{L^p(0,\infty;X)}\big)$$
$$\leq c(\|f\|_{L^p(0,\infty;X)} + \|u_0\|_{\mathscr{F}_A^p}).$$

Finally, due to the norm equivalence of $\mathscr{F}_A^p$ and the real interpolation space, we obtain the result. $\qquad\square$

Remark 1.3.6. (i) Note that, of course, the canonical space of initial values is the set of traces $\mathscr{F}_A^p$. If we consider any $u_0 \in (X, \mathcal{D}(A))_{1-\frac{1}{p}, p}$, then we proved in Lemma 1.3.3 that $u_0 \in \mathscr{F}_{A,\infty}^p \subset \mathscr{F}_{A,T}^p$ for every $0 < T < \infty$. Furthermore, we obtain

$$\lim_{T \to 0} \|u_0\|_{\mathscr{F}_{A,T}^p} = 0$$

due to the definition of $W_{A,T}^p$ in (1.10).

(ii) If we endow the space of initial values with the norm of $\mathscr{F}_A^p$, then the constant c_i in the maximal L^p-regularity estimate with initial value $u_0 \neq 0$ can be estimated by $c_i \leq 2c_h + 1$, where c_h denotes the constant in the maximal L^p-regularity estimate with initial value $u_0 = 0$.

One of the main ingredients to prove a maximal regularity result for a non-autonomous system, which will be discussed in Chapter 2, is a uniformity for a family of operators in the estimate (1.8). The main tool to prove this will be the next lemma.

Lemma 1.3.7. *Let X be a Banach space and A, B be operators such that $\mathcal{D}(A) = \mathcal{D}(B)$. Let $c_A > 0$, assume that $A \in \mathcal{MR}_p(X, c_A)$, and $B \in \mathcal{MR}_p(X)$. Finally, assume that*

$$\|(A - B)x\|_X \leq \frac{1}{2c_A}\|Ax\|_X$$

holds true for all $x \in \mathcal{D}(A)$. Then $B \in \mathcal{MR}_p(X, 2c_A + 1)$.

Remark 1.3.8. Note that Lemma 1.3.7 deals only with the case of initial value $u_0 = 0$. As discussed in Remark 1.3.6 (ii) this implies that the constant in the maximal L^p-estimate with initial value $u_0 \neq 0$ of the operator B can be estimated by $4c_A + 3$ if we endow the space of initial values with the norm corresponding to $\mathscr{F}_B^p$.

Proof of Lemma 1.3.7. Let $f \in L^p(0, \infty; X)$. Then, since B has maximal L^p-regularity, we obtain the existence of a solution $u \in W_B^p$ to the abstract Cauchy problem with data f and the operator B. Thus, it holds that

$$\partial_t u + Au = f + (A - B)u, \ u(0) = 0,$$

and hence we obtain the estimate

$$\|\partial_t u\|_{X,p} + \|Au\|_{X,p} \leq c_A\Big(\|f\|_{X,p} + \|(A-B)u\|_{X,p}\Big)$$
$$\leq c_A\Big(\|f\|_{X,p} + \frac{1}{2c_A}\|Au\|_{X,p}\Big)$$

holds true, and the last addend can be absorbed on the left-hand side. Thus we can estimate as follows

$$\|\partial_t u\|_{X,p} + \|Bu\|_{X,p} \leq \|\partial_t u\|_{X,p} + \|(A-B)u\|_{X,p} + \|Au\|_{X,p}$$
$$\leq \|\partial_t u\|_{X,p} + \Big(1 + \frac{1}{2c_A}\Big)\|Au\|_{X,p}$$
$$\leq (2c_A + 1)\|f\|_{X,p},$$

and this implies $B \in \mathcal{MR}_p(X, 2c_A + 1)$. □

Note that to prove the perturbation argument in Lemma 1.3.7, we already needed to assume that $B \in \mathcal{MR}_p(X)$. In the last part of this section we will present an equivalent formulation of maximal L^p-regularity due to L. Weis and present a perturbation result to obtain maximal L^p-regularity.

Definition 1.3.9. Let X, Y be Banach spaces and $\mathcal{T} \subset \mathcal{L}(X, Y)$ be a family of operators. The set $\mathcal{T}$ is called $\mathcal{R}$-*bounded* if there exist $c > 0$ and $p \in [1, \infty)$ such that for all $N \in \mathbb{N}$, $T_j \in \mathcal{T}$ and $x_j \in X$, and all symmetric and independent $\{-1, 1\}$-valued random variables ε_j on some probability space $(\Omega, \mathcal{M}, \mu)$, $1 \leq j \leq N$, the inequality

$$\Big\|\sum_{j=1}^{N} \varepsilon_j T_j x_j\Big\|_{L^p(\Omega, Y)} \leq c\Big\|\sum_{j=1}^{N} \varepsilon_j x_j\Big\|_{L^p(\Omega, X)}$$

holds true. The smallest of such constant c is called the $\mathcal{R}$-*bound* of $\mathcal{T}$ and will be denoted by $\mathcal{R}(\mathcal{T})$.

The notion of $\mathcal{R}$-boundedness is obviously stronger than uniform boundedness in the operator topology. This can be seen by choosing $N = 1$ in the definition above. If X and Y are Hilbert spaces, one may choose $p = 2$ and one can see easily that uniform boundedness implies $\mathcal{R}$-boundedness, so that both notions are equivalent in this setting.

Definition 1.3.10. Let X be a Banach space. A sectorial operator A on the Banach space X is called $\mathcal{R}$-*sectorial*, if

$$\mathcal{R}(\{\lambda(\lambda + A)^{-1} \mid \lambda > 0\}) < \infty.$$

Furthermore, the $\mathcal{R}$-angle $\phi_A^{\mathcal{R}}$ of an $\mathcal{R}$-sectorial operator A is defined by

$$\phi_A^{\mathcal{R}} := \inf\left\{\theta \in (0, \pi) \mid \mathcal{R}(\{\lambda(\lambda + A)^{-1} \mid |\arg(\lambda)| < \pi - \theta\}) < \infty\right\}.$$

Next, let us state a perturbation result for $\mathcal{R}$-sectorial operators, see [10, 4.2 Proposition].

Lemma 1.3.11. *Let X be a Banach space and A be an $\mathcal{R}$-sectorial operator. Furthermore, let B be a linear operator on X such that $\mathcal{D}(A) \subset \mathcal{D}(B)$ and $\|Bx\| \leq \alpha\|Ax\|$ for some $\alpha > 0$ and all $x \in \mathcal{D}(A)$. Assume furthermore that*

$$\mathcal{R}(\{A(\lambda + A)^{-1} \mid \lambda \in S_\theta\}) = a < \infty$$

for some $\theta > 0$. Then

$$\mathcal{R}(\{\lambda(\lambda + A + B)^{-1} \mid \lambda \in S_\theta\}) < \infty$$

provided that $\alpha < \frac{1}{a}$.

Hence the previous lemma implies that the notion of $\mathcal{R}$-boundedness is stable under small perturbations. Finally, the last result we will present in this section shows that the notion of $\mathcal{R}$-boundedness yields a useful characterisation of maximal L^p-regularity, see [10, 4.4 Theorem].

Definition 1.3.12. A Banach space X is said to be of class $\mathcal{HT}$ if the X-valued *Hilbert transform*

$$f \mapsto \mathrm{p.v.} \int_{-\infty}^{\infty} \frac{f(s)}{t - s} \mathrm{d}s, \quad f \in C_0^\infty(\mathbb{R}; X),$$

extends to a bounded operator on $\mathrm{L}^p(\mathbb{R}; X)$ for some $1 < p < \infty$. Here, p.v. denotes the principle value.

It is well-known, see [29, Ch. 4], that $\mathbb{R}^n$ and $\mathbb{C}^n$ are of class $\mathcal{HT}$. Furthermore, if X is of class $\mathcal{HT}$, then the Lebesgue–Bochner space $L^p(I; X)$ for some interval $I \subset \mathbb{R}$ and $1 < p < \infty$ is also of class $\mathcal{HT}$. A proof for this can be found in [1, Theorem 4.5.2].

In the 1980's , J. Bourgain [7] and D. L. Burkholder [8] proved that a Banach space is of class $\mathcal{HT}$ if and only if X is a $\mathcal{UMD}$-space and if and only if X is ζ-convex.

Proposition 1.3.13. *Let X be a Banach space of class $\mathcal{HT}$ and A be a sectorial operator with spectral angle $\phi_A < \frac{\pi}{2}$. Then the following statements are equivalent:*

(i) *The operator A has maximal L^p-regularity.*

(ii) *The set $\{A(\lambda + A)^{-1} \mid \lambda \in S_\theta\}$ is $\mathcal{R}$-bounded for some $\theta > \frac{\pi}{2}$.*

(iii) *The set $\{A(i\lambda + A)^{-1} \mid \lambda \in \mathbb{R}\}$ is $\mathcal{R}$-bounded.*

Combining the last results, we obtain the following.

Proposition 1.3.14. *Let X be a Banach space of class $\mathcal{HT}$, let $c_A > 0$ and let $A \in \mathcal{MR}_p(X, c_A)$. Then there exists $\varepsilon > 0$ with the following property: Assume that B is a sectorial operator with $\mathcal{D}(A) = \mathcal{D}(B)$ and $\|(A - B)x\|_X \le \varepsilon \|Ax\|_X$ for all $x \in \mathcal{D}(A)$. Then $B \in \mathcal{MR}_p(X, 2c_A + 1)$.*

Proof. Since X is a Banach space of class $\mathcal{HT}$, we obtain that A is $\mathcal{R}$-sectorial in a sector with opening angle larger than $\frac{\pi}{2}$. Using Lemma 1.3.11 with the operators A and $B - A$, we obtain that also the $\lambda(\lambda + B)^{-1}$ is $\mathcal{R}$-bounded on a sector, provided that ε is small enough. Hence, using the characterization in Proposition 1.3.13 once again, we obtain that B has maximal L^p-regularity. Finally, Lemma 1.3.7 yields the desired result provided that ε is small enough. $\qquad\square$

Let us finally state a corollary of Proposition 1.3.13.

Corollary 1.3.15. *Let X be a Banach space of class $\mathcal{HT}$ and A be a sectorial operator with spectral angle $\phi_A < \frac{\pi}{2}$. Furthermore, let A fulfil one and thus all statements of Proposition 1.3.13. Then for every $\mu > 0$ the operator $A + \mu$ has maximal L^p-regularity.*

Proof. The convexity of $\mathcal{R}$-bounds implies

$$\mathcal{R}\{\lambda(\lambda + \mu + A)^{-1} \mid \lambda \in S_\theta\} \leq \mathcal{R}\{\lambda(\lambda + A)^{-1} \mid \lambda \in S_\theta\},$$

see the proof of Proposition 4.3 in [10]. Since the $\mathcal{R}$-boundedness of

$$\mathcal{R}\{\lambda(\lambda + \mu + A)^{-1} \mid \lambda \in S_\theta\}$$

is equivalent to the $\mathcal{R}$-boundedness of

$$\mathcal{R}\{(A + \mu)(\lambda + \mu + A)^{-1} \mid \lambda \in S_\theta\}$$

since

$$(A + \mu)(\lambda + \mu + A)^{-1} = \mathrm{Id} - \lambda(\lambda + \mu + A)^{-1},$$

and using [10, 3.4 Proposition], we obtain that $A + \mu$ has maximal L^p-regularity. $\square$

In the situation of the last corollary, let $\mu > 0$, $f \in \mathrm{L}^p(0, T; X)$ for some finite $T \in (0, \infty)$, and $(\partial_t + A)u = f$. Then we obtain with $\tilde{f}(t) = e^{-\mu t}f(t)$ and $v(t) = e^{-\mu t}u(t)$ that

$$
\begin{aligned}
\|\mu u\|_{\mathrm{L}^p(0,T;X)} &\leq c(\mu, T)\|\mu v\|_{\mathrm{L}^p(0,T;X)} \\
&\leq c(\mu, T)\|(\mu + A)v\|_{\mathrm{L}^p(0,T;X)} \\
&\leq c(\mu, T)\left(\|\tilde{f}\|_{\mathrm{L}^p(0,T;X)} + \|u(0)\|_{\mathscr{I}^p_{A+\mu}}\right) \\
&\leq c(\mu, T)\left(\|f\|_{\mathrm{L}^p(0,T;X)} + \|u(0)\|_{\mathscr{I}^p_{A+1}}\right).
\end{aligned}
$$

(1.13)

Here, we used the resolvent estimate, and the maximal L^p-regularity of $A + \mu$. We will use this estimate frequently in Chapter 2.

Remark 1.3.16. In the situation of the last corollary, we proved that if A has maximal L^p-regularity, then $A + \mu$ has maximal L^p-regularity. But the set of initial values might differ. The set of initial values for the initial value problem to the operator A is the set of all traces $u_0 = u(0)$ to functions u such that $\partial_t u, Au \in \mathrm{L}^p(0, T; X)$. If we consider $A + \mu$, we obtain the condition that $\partial_t u, (A + \mu)u \in \mathrm{L}^p(0, T; X)$, which is equivalent to the assumption that $\partial_t u, Au, \mu u \in \mathrm{L}^p(0, T; X)$.

In the subsequent chapters, we will frequently consider the shifted operator $A + \mu$ on a finite time interval. If we do so, we have to change the norm of the initial value on the right-hand side of the maximal L^p-regularity estimate to the norm $\|u_0\|_{\mathscr{I}^p_{A+1}}$ or one of the equivalent norms to this.

1.4 The Sum of Closed Operators

In this section we would like to prepare the chapter about the strong solvability of the Navier–Stokes equations under appropriate assumptions on the data. As usual in the theory of non-linear partial differential equations, it will be one of the important steps to prove that the non-linear part in the $L^p(L^q)$-norm can be estimated against the solution with respect to the norm of the solution space in a suitable way. The main ingredient of this proof will be the following Theorem 1.4.1. This so-called *Mixed Derivative Theorem* goes back to P.E. Sobolevskii [49]; here we will cite the version of J. Escher, J. Prüss, and G. Simonett [13].

Theorem 1.4.1 (Mixed Derivative Theorem). *Suppose that A and B are commuting sectorial operators in a Banach space X with the following property: For all $\lambda > 0$ the operator $A + \lambda B$ is closed in the domain $\mathcal{D}(A + \lambda B) := \mathcal{D}(A) \cap \mathcal{D}(B)$. Furthermore, there exists $c > 0$ such that*

$$(1.14) \qquad \|Ax\|_X + \lambda\|Bx\|_X \leq c\|(A + \lambda B)x\|_X$$

holds for all $x \in \mathcal{D}(A) \cap \mathcal{D}(B)$. Then there exists a constant $C > 0$ such that the estimate

$$\|A^\alpha B^{1-\alpha}x\|_X \leq C\|Ax + Bx\|_X$$

holds for all $x \in \mathcal{D}(A) \cap \mathcal{D}(B)$ and all $\alpha \in (0, 1)$.

Usually, this theorem is applied to differential operators A, which only acts on the spatial variables, and the time derivative. Hence, it is obvious, that those operators commute. Usually, the crucial part is to prove the reverse triangle inequality (1.14). The first who proved a very general theorem which gives a sufficient condition such that (1.14) holds were G. Dore and A. Venni [11] and J. Prüss and H. Sohr [43].

Theorem 1.4.2. *Let X be a Banach space of class $\mathcal{HT}$. Let $A, B \in \mathcal{BIP}(X)$ be resolvent commuting operators with power angles θ_A and θ_B and assume $\theta_A + \theta_B < \pi$. Then the operator $A + B$ with natural domain $\mathcal{D}(A) \cap \mathcal{D}(B)$ is closed, and there exists $c > 0$ such that*

$$\|Ax\|_X + \|Bx\|_X \leq c\|(A + B)x\|_X$$

holds for all $x \in \mathcal{D}(A) \cap \mathcal{D}(B)$.

Note that we cited the version by J. Prüss and H. Sohr [43, Theorem 4]. It is more general then the version by G. Dore and A. Venni Theorem since they assumed that $0 \in \rho(A) \cap \rho(B)$. To prove that we can apply the Mixed Derivative Theorem to operators A and B which fulfil the assumptions of Theorem 1.4.2 it remains to prove that the constant in (1.14) can be chosen uniformly in λ. This can be done by checking the proof of Theorem 1.4.2. Since we want to avoid to do this, let us mention that the theory of N. J. Kalton and L. Weis [34] yields a different proof of the Mixed Derivative Theorem. They proved, that if the operators A and B commute, A has a bounded $\mathcal{H}^\infty$-caluculus, and B is $\mathcal{R}$-sectorial such that $\phi_A^\infty + \phi_B^{\mathcal{R}} < \pi$, then the statement of the Mixed Derivative Theorem holds for the operators A and B, see [42, Corollary 4.5.10].

Theorem 1.4.3. *Let X be a Banach space of class $\mathcal{HT}$, let $T \in (0, \infty)$, and let $p \in (1, \infty)$. Let us consider the time derivative*

$$\partial_t \colon \{u \in W^{1,p}(0,T;X) \mid u(0) = 0\} \to \mathrm{L}^p(0,T;X),$$
$$u \mapsto \partial_t u.$$

Then ∂_t is an invertible operator. Furthermore, we have that ∂_t is of class $\mathcal{BIP}$ with power angle $\theta_{\partial_t} \leq \frac{\pi}{2}$. Moreover, the estimate

$$\|\partial_t^{i\alpha}\|_{\mathcal{L}(\mathrm{L}^p(0,T;X))} \leq (1 + |\alpha|^2) e^{\frac{\pi}{2}|\alpha|}$$

holds true for all $\alpha \in \mathbb{R}$.

Note that since the right-hand side in the estimate of the operator norm of $\partial_t^{i\alpha}$ does not depend on T, the same also holds true for the time derivative if we extend the time interval to $[0, \infty)$. Hence the restriction $T < \infty$ is not necessary. Let us finally combine the last results in the version we will finally make use of it.

Theorem 1.4.4. *Let X be a Banach space of class $\mathcal{HT}$, let $1 < p < \infty$, and let $T \in (0, \infty]$. Let $A \in \mathcal{BIP}(X)$ be an operator with power angle $\theta_A < \frac{\pi}{2}$. Furthermore, let $\tilde{A}$ denote its extension onto $\mathrm{L}^p(0,T;X)$ by*

$$(\tilde{A}u)(t, \cdot) = A(u(t, \cdot)).$$

Then $\partial_t + \tilde{A}$ is a closed operator, the estimate (1.14) with $B = \partial_t$ and $A = \tilde{A}$ holds true for all $x \in \mathcal{D}(\tilde{A}) \cap \dot{W}^{1,p}(0, T; X)$, and we have

$$\|\tilde{A}^\alpha \partial_t^{1-\alpha} u\| \leq c \big(\|\tilde{A}u\| + \|\partial_t u\| \big),$$

where the norm $\|\cdot\|$ denotes the $\mathrm{L}^p(0, T; X)$ norm.

Proof. Since A has $\mathcal{BIP}$ we trivially obtain that $\tilde{A}$ has $\mathcal{BIP}$ and the power angle remains unchanged. Hence we can apply the Dore–Venni Theorem as well as the Mixed Derivative Theorem to obtain the result. $\square$

We will apply the Mixed Derivative Theorem in Chapter 3 to obtain an estimate for the convective term in the Navier–Stokes equations. Crucial to obtain that estimate is a Sobolev embedding for fractional derivatives. More precisely, we need the following proposition.

Proposition 1.4.5. *Let X be a Banach space of class $\mathcal{HT}$, let $T \in (0, \infty]$, and let $\alpha \in (0, 1)$. Furthermore, let $p < \frac{1}{\alpha}$, let $q = (\frac{1}{p} - \alpha)^{-1}$, and let us consider the time derivative*

$$\partial_t \colon \{u \in W^{1,p}(0, T; X) \mid u(0) = 0\} \to \mathrm{L}^p(0, T; X),$$
$$u \mapsto \partial_t u.$$

Then there exists $c > 0$ such that

$$\|u\|_{\mathrm{L}^q(0,T;X)} \leq c \|\partial_t^\alpha u\|_{\mathrm{L}^p(0,T;X)}$$

holds true for all $u \in \mathcal{D}(\partial_t^\alpha)$.

This result seems to be well-known since it is used in several papers, see for instance [46]. Nevertheless, we did not find a proof of this result in the literature, and therefore we present a proof here.

Proof. Since ∂_t has $\mathcal{BIP}$, we obtain that

$$\mathcal{D}(\partial_t^\alpha) = [\mathrm{L}^p(0, T; X); W_{(0)}^{1,p}(0, T; X)]_\alpha$$

due to Proposition 1.2.4. Here, the subset (0) indicates zero trace in $t = 0$.

Note that the linear hull of functions of type $\varphi(t)x$ with $\mathbb{R}$-valued $W_{(0)}^{1,p}$-functions φ and $x \in X$ are dense in $L^p(0,T;X)$ and $W_{(0)}^{1,p}(0,T;X)$, and hence it suffices to consider the case of $\mathbb{R}$-valued functions.

Finally, by using an extension argument, we can restrict ourselves to the case $T = \infty$, and by using an odd extension, we can assume that the time axis is equal to $\mathbb{R}$.

We obtain that

$$[L^p(\mathbb{R}); W^{1,p}(\mathbb{R})]_\alpha = H^{\alpha,p}(\mathbb{R}) = F_{p,2}^\alpha(\mathbb{R}),$$

where $H^{\alpha,p}$ denotes a Bessel potential space and $F_{p,2}^\alpha$ a Triebel–Lizorkin space. For the first equality we refer to [37, Example 2.12] and for the second to [53, Section 2.3.3]. Also, the definition of those spaces can be found in the cited monographs.

Due to [53, Theorem 2.8.1], we obtain the embedding

$$F_{p,2}^\alpha(\mathbb{R}) \hookrightarrow L^q(\mathbb{R}).$$

So far, we proved the existence of $c > 0$ such that

(1.15) $$\|u\|_{L^q(0,T;X)} \le c\big(\|u\|_{L^p(0,T;X)} + \|\partial_t^\alpha u\|_{L^p(0,T;X)}\big)$$

holds true and we have to prove that we can neglect the first term on the right-hand side. If T is finite, we can use Poincaré's inequality to obtain that

$$\|u\|_{L^p(0,T;X)} \le c_T \|\partial_t u\|_{L^p(0,T;X)},$$

what implies

$$\|u\|_{L^p(0,T;X)} \le c_T \|\partial_t^\alpha u\|_{L^p(0,T;X)}.$$

Otherwise, if $T = \infty$, let us argue that for every $\mu > 0$ we have

$$\|\partial_t^\alpha u(\mu \cdot)\|_{L^p(0,\infty;X)} = \mu^{\alpha - \frac{1}{p}} \|\partial_t^\alpha u\|_{L^p(0,\infty;X)}.$$

This can be seen as follows. Assume first that

$$u \in E := \mathcal{D}(\partial_t) \cap \operatorname{Im}(\partial_t).$$

Note that E is a dense subspace of $W^{1,p}_{(0)}(0, \infty; X)$ and let us consider the operator $T_\mu \colon u \mapsto u(\mu \cdot)$, which maps E into E. Then we have

$$\partial_t^\alpha (T_\mu u) = \frac{\sin(\alpha \pi)}{\pi} \int_0^\infty \lambda^{\alpha-1} (\lambda + \partial_t)^{-1} \partial_t (T_\mu u) \, \mathrm{d}\lambda,$$

see the proof of [10, 2.3 Theorem] on page 21. This implies

$$\|\partial_t^\alpha (T_\mu u)\|^p_{\mathrm{L}^p(0,\infty;X)}$$
$$= \Big(\frac{\sin(\alpha \pi)}{\pi}\Big)^p \int_0^\infty \Big| \int_0^\infty \lambda^{\alpha-1} \int_0^t e^{-\lambda(t-s)} \partial_s u(\mu s) \, \mathrm{d}s \, \mathrm{d}\lambda \Big|^p \mathrm{d}t.$$

Let us substitute $s' = \mu s$, $t' = \mu^{-1} t$, and $\lambda' = \mu^{-1}\lambda$ and we obtain

$$\|\partial_t^\alpha (T_\mu u)\|^p_{\mathrm{L}^p(0,\infty;X)}$$
$$= \Big(\mu^{\alpha - \frac{1}{p}} \frac{\sin(\alpha \pi)}{\pi}\Big)^p \int_0^\infty \Big| \int_0^\infty (\lambda')^{\alpha-1} \int_0^t e^{-\lambda'(t'-s')} \partial_{s'} u(s') \, \mathrm{d}s' \, \mathrm{d}\lambda' \Big|^p \mathrm{d}t'$$
$$= \Big(\mu^{\alpha - \frac{1}{p}}\Big)^p \|\partial_t^\alpha u\|^p_{\mathrm{L}^p(0,\infty;X)}$$

This proves that the term on the left-hand side of (1.15) scales on the dense subspace as $\mu^{-\frac{1}{q}} = \mu^{\alpha - \frac{1}{p}}$, and due to a density argument, this scaling property holds for all $u \in \mathcal{D}(\partial_t^\alpha)$. The second term on the right-hand side has the same scaling property, but the first scales as $\mu^{-\frac{1}{p}}$. Hence, considering the limit $\mu \to \infty$, we obtain that the first term on the right-hand side of (1.15) can be neglected. $\qquad \square$

1.5 Stokes Operator

In this final section of the preliminaries we would like to introduce the *Stokes operator* and the *Helmholtz projection*. For this purpose, let $\Omega \subset \mathbb{R}^n$ denote a domain and let $f \in \mathrm{L}^q(\Omega)$, $1 < q < \infty$.

Definition 1.5.1. Let $1 < q < \infty$ and let $\Omega \subset \mathbb{R}^n$ be a domain. We say that there exists an L^q-*Helmholtz decomposition* on Ω if for all $f \in \mathrm{L}^q(\Omega)$ there exists $f_0 \in \mathrm{L}^q_\sigma(\Omega)$ and $p \in \dot{W}^{1,q}(\Omega)$ such that

$$(1.16) \qquad\qquad\qquad f = f_0 + \nabla p,$$
$$\|f_0\|_q + \|\nabla p\|_q \le c\|f\|_q,$$

with a constant $c > 0$ which does not depend on f, and the addends of this composition are unique. In this case, the projection

$$P_{q,\Omega} \colon f \to f_0$$

is called the L^q-*Helmholtz projection* on Ω.

By formally applying the divergence to (1.16), or to be more precise by multiplying with $\nabla\varphi$ for some function $\varphi \in C^\infty(\Omega)$ and integrating over Ω, we see that the pressure p is a solution to the *weak Neumann problem*

(1.17)
$$\Delta p = \operatorname{div} f \text{ in } \Omega,$$
$$n \cdot \nabla p = f \cdot n \text{ on } \partial\Omega,$$

where n denotes the outer normal vector. We call $p \in \dot{W}^{1,q}(\Omega)$ a solution to (1.17), if

$$\int_\Omega (\nabla p(x) - f(x)) \cdot \nabla\varphi(x)\mathrm{d}x = 0$$

for all $\varphi \in \dot{W}^{1,q'}(\Omega)$, where q' denotes the conjugate exponent to q. Hence we get that there exists an L^q-Helmholtz decomposition on Ω if and only if the weak Neumann problem (1.17) is *uniquely* solvable. Here, uniquely solvable means that the pressure p is determined up to an additive constant. Due to the *Lemma of Lax–Milgram* for all domains $\Omega \subset \mathbb{R}^n$ there exists an L^2-Helmholtz decomposition. It is known, see [22, Remark III.1.3], that there exist domains with smooth, but non compact boundary, such that the Helmholtz decomposition fails. Nowadays, there is a zoo of papers in which the existence of a Helmholtz projection in several kinds of domains is proved, *e.g.*, in bounded or exterior domains, in whole or half-space or in domains with non-compact boundaries. Important examples are the works by D. Fujiwara and H. Morimoto [21] (bounded domains with smooth boundary), T. Miyakawa [39] (exterior domains with smooth boundary), and R. Farwig and H. Sohr [20] (aperture domains).

In 1992, C. Simader and H. Sohr [48] generalised those results to all domains with compact C^1 boundary.

It is easy to see that if $p \in \dot{W}^{1,q}(\Omega) \cap \dot{W}^{1,r}(\Omega)$, $q \neq r$, is a $\dot{W}^{1,q}$-solution to (1.17), then p is also a $\dot{W}^{1,r}$-solution. Hence on the set $L^q(\Omega) \cap L^r(\Omega)$, the Helmholtz projections P_q and P_r coincide. Therefore, it seems to be suitable to omit the index q.

Note that if Ω is a domain with compact and $C^{1,1}$-smooth boundary, the domain of the Dirichlet Laplacian is known to be equal to

$$\mathcal{D}(\Delta) = W^{2,q}(\Omega) \cap W_0^{1,q}(\Omega),$$

see [27, Chapter 9]. Hence we can define the Stokes operator $A_{q,\Omega}$ as follows. In the following, we will consider only the case of domains in $\mathbb{R}^3$.

Definition 1.5.2. Let $\Omega \subset \mathbb{R}^3$ be the half-space, $\mathbb{R}^3$ or a domain with compact $C^{1,1}$-smooth boundary. Then we define the Stokes operator

$$A_{q,\Omega} \colon \mathcal{D}(A_{q,\Omega}) \to L^q_\sigma(\Omega), \ u \mapsto -P_{q,\Omega} \Delta u$$

with

$$\mathcal{D}(A_{q,\Omega}) := W^{2,q}(\Omega) \cap W_0^{1,q}(\Omega) \cap L^q_\sigma(\Omega).$$

If there is no confusion, we will omit the index q or Ω.

Let us now assume that $\Omega \subset \mathbb{R}^3$ is a bounded or exterior domain with $C^{1,1}$-boundary.

Obviously, A_q is a densely defined operator and, since the adjoint operator to P_q is given by $P_{q'}$, $\frac{1}{q} + \frac{1}{q'} = 1$, we have that $(A_q)' = A_{q'}$ and hence A_q has dense range. Furthermore, the Stokes operator fulfils the *resolvent estimate*

$$(1.18) \qquad |\lambda| \|u\|_q + \|\nabla^2 u\|_q + \|\nabla p\|_q \le c(\lambda) \|f\|_q, \ \lambda \in \mathbb{C} \setminus (-\infty, 0],$$

where (u, p) is a solution to

$$\lambda u - \Delta u + \nabla p = f, \ \operatorname{div} u = 0,$$

and endowed with the homogeneous Dirichlet boundary condition

$$u_{|\partial \Omega} = 0,$$

see [24, Theorem 1] for the case of a bounded domain with smooth boundary, or [6] if Ω is an exterior domain with $C^{2+\mu}$-boundary, $0 < \mu < 1$. For the case of a $C^{1,1}$-boundary we refer to [19].

The constant c in (1.18) usually depends on the domain, on q and on λ. To be more precise, the constant can be chosen uniformly in each sector

$$S_{\delta,\varepsilon} := \{\lambda \in \mathbb{C} \setminus \{0\} \mid |\arg(\lambda)| < \pi - \varepsilon\} \setminus B_\delta(0).$$

Moreover, as proved in [19], the constant in (1.18) can be chosen uniformly in δ if Ω is a bounded domain, or if Ω is an exterior domain and $q < \frac{3}{2}$. If Ω is bounded, we further obtain that the Stokes operator is boundedly invertible. Note that the uniformity of (1.18) implies that

$$(1.19) \qquad \|\nabla^2 u\|_q \leq c\|Au\|_q, \; 1 < q < \frac{3}{2},$$

for all $u \in \mathcal{D}(A)$. To prove this estimate, we consider the limit $\lambda \to 0$ in (1.18). Note that the converse estimate „$\geq$" in (1.19) follows trivially by the continuity of the Helmholtz projection.

Due to $A_{q'} = (A_q)'$ and the uniformity of the resolvent estimate in δ for $q < \frac{3}{2}$ we obtain that

$$\|(\lambda + A_q)^{-1}\|_{\mathcal{L}(\mathrm{L}^q_\sigma(\Omega))} \leq \frac{c}{|\lambda|}$$

for all $q > 3$ and, due to an interpolation result, even for all $1 < q < \infty$. Hence, the Stokes operator is generator of a bounded analytic semigroup denoted by $\left(e^{-tA_q}\right)_{t \geq 0}$.

Note that the estimate

$$(1.20) \qquad \|\nabla u\|_q \leq c\|A^{\frac{1}{2}}u\|_q$$

hold true for all exterior domains $\Omega \subset \mathbb{R}^3$ with sufficiently smooth boundary and for $q < 3$, see [4, Theorem 4.4]. The converse estimate holds true for all $q \in (1, \infty)$.

The estimate (1.20) enables us to prove the continuity of the operator $A^{-\frac{1}{2}}P\mathrm{div}$, which is important in the theory of weak solutions to the Navier–Stokes equations. Let $u \in C_0^\infty(\Omega)$ and $w \in \mathcal{D}(A_{q'}) \cap \mathcal{R}(A_{q'})$, where

q' denotes the conjugate exponent to q, hence $\frac{1}{q} + \frac{1}{q'} = 1$. Then we have

$$
\begin{aligned}
|\langle A_q^{-\frac{1}{2}} P_q \operatorname{div} u, w\rangle| &= |\langle P_q \operatorname{div} u, A_{q'}^{-\frac{1}{2}} w\rangle| \\
&= |\langle \operatorname{div} u, P_{q'} A_{q'}^{-\frac{1}{2}} w\rangle| \\
&= |\langle u, \nabla A_{q'}^{-\frac{1}{2}} w\rangle| \\
&\leq \|u\|_q \|\nabla A_{q'}^{-\frac{1}{2}} w\|_{q'} \\
&\leq \|u\|_q \|A_{q'}^{\frac{1}{2}} A_{q'}^{-\frac{1}{2}} w\|_{q'} \\
&\leq \|u\|_q \|w\|_{q'}
\end{aligned}
$$

if $q' < 3$, or equivalently $q > \frac{3}{2}$. Hence we obtain

$$
\text{(1.21)} \qquad A^{-\frac{1}{2}} P \operatorname{div} \in \mathcal{L}(\mathrm{L}^q(\Omega)), \ q > \frac{3}{2}.
$$

Since A_q is a sectorial operator, the calculus described in Section 1.2 ensures us to define A_q^α for all $\alpha \geq 0$, and due to [4, Proposition 3.2] we obtain that A_q^α is injective. Moreover, since A_q is the generator of a bounded analytic semigroup, we obtain that

$$
\sup \left\{ \|t A_q e^{-t A_q}\|_{\mathcal{L}(\mathrm{L}^q_\sigma)} \mid t > 0 \right\}
$$

is uniformly bounded. Thus, an application of the *moment inequality*

$$
\|A^\alpha x\| \leq c \|x\|^\alpha \|Ax\|^{1-\alpha}
$$

for $x \in \mathcal{D}(A)$ yields

$$
\text{(1.22)} \qquad \|A_q^\alpha e^{-t A_q} u\|_q \leq c t^{-\alpha} \|u\|_q
$$

for all $u \in \mathrm{L}^q_\sigma(\Omega)$ and all $1 < q < \infty$. Here, the constant c may depend on Ω and q, but is independent of u. In addition, W. Borchers and T. Miyakawa also proved in [4, Corollary 4.5] that

$$
\text{(1.23)} \qquad \|u\|_r \leq c \|A_q^\alpha u\|_q
$$

for all $u \in \mathcal{D}(A_q)$ and

$$1 < q < 3, \quad 0 < \alpha < \frac{3}{2q}, \quad \frac{1}{q} = \frac{1}{r} + \frac{2\alpha}{3}.$$

Combining the inequalities (1.22), (1.23), and (1.20) we obtain that

(1.24)
$$\|e^{-tA_q}v\|_r \leq ct^{-\frac{3}{2}\left(\frac{1}{q}-\frac{1}{r}\right)}\|v\|_q, \quad 0 \leq \alpha \leq 1, t > 0, r \geq q$$
$$\|\nabla e^{-tA_q}v\|_q \leq ct^{-\frac{1}{2}-\frac{3}{2}\left(\frac{1}{r}-\frac{1}{q}\right)}\|v\|_r, \quad 1 < r \leq q \leq 3.$$

Note that the constraint $q < 3$ as in (1.23) is not needed in (1.24)$_2$ by a famous result by H. Iwashita [33].

Y. Giga and H. Sohr proved in [24] and [25] that the Stokes operator has $\mathcal{BIP}$ and its power angle is 0. As discussed in Section 1.3, this implies that the Stokes operator has maximal L^p-regularity. Finally, let us mention that A. Noll and J. Saal proved that the Stokes operator has also a bounded $\mathcal{H}^\infty$-calculus if $\partial\Omega$ is of class C^3, see [41].

CHAPTER 2

Non-Autonomous Stokes System

In this chapter we are going to prove one of the main results of this dissertation. As discussed in the preface, we would like to prove a maximal regularity result for the Stokes equations in a non-cylindrical time-space domain.

Let us start by stating the main result of this chapter. For $t \in [0, \infty)$ let $\Omega(t)$ denote an exterior domain with $\partial\Omega(t) \in C^3$. The non-cylindrical time-space domain Q is defined by

$$Q := \bigcup_{t>0} \left(\Omega(t) \times \{t\} \right).$$

Furthermore, let

$$\Gamma := \bigcup_{t>0} \left(\partial\Omega(t) \times \{t\} \right)$$

denote the boundary corresponding to the spatial variable x. We would like to prove a maximal regularity result for the system

$$
\begin{aligned}
v_t - \Delta v + \nabla p &= f \quad \text{in } Q, \\
\operatorname{div} v &= 0 \quad \text{in } Q, \\
v &= 0 \quad \text{on } \Gamma, \\
v(0) &= v_0 \quad \text{in } \Omega(0) =: \Omega_0.
\end{aligned}
$$

(2.1)

In the following, we assume that the evolution of the domain is known *a priori*. Before we state the assumption on the movement of the domain carefully, let us introduce the space

$$C_b^{3,1} := \{f \in C^0(\Omega_0 \times (0,\infty); \mathbb{R}^3 \times (0,\infty)) \mid \partial_t^k \partial_\xi^\alpha f \in C^0,$$
$$1 \le 2k + |\alpha| \le 3, \; k \in \mathbb{N}_0, \; \alpha \in \mathbb{N}_0^3\}.$$

To $\psi \in C_b^{3,1}$ such that $\psi(\xi,t) = (\phi(\xi,t),t)$ for a suitable function ϕ, let us introduce the (semi-)norm

(2.2)
$$|\phi(t_0)|_{C_b^{3,1}} := \|\phi(t_0)\|_{C^3(\Omega_0)} + \|\phi_t(t_0)\|_{C^1(\Omega_0)},$$
$$\|\phi\|_{C_b^{3,1}} := \sup_{t_0 \in [0,\infty)} |\phi(t_0)|_{C_b^{3,1}}.$$

The assumptions on the movement of the domain read as follows.

Assumption 2.0.1. *Let us assume that there exists a family of exterior domains $\Omega(t) \subset \mathbb{R}^3$ with $\partial\Omega \in C^3$ and a map*

$$\psi \colon \overline{\Omega_0 \times (0,\infty)} \to \overline{Q}, \; (\xi,t) \mapsto \psi(\xi,t) =: (\phi(\xi,t),t).$$

Furthermore, we assume that the function ϕ has the following properties.

(i) *For $t \in [0,\infty)$ the map $\phi(\cdot,t) \colon \overline{\Omega_0} \to \overline{\Omega(t)}$ is a C^3-diffeomorphism; its inverse (for fixed t) will be denoted by $\phi(\cdot,t)^{-1}$. Moreover, let $\phi(\cdot,0) = \text{Id}$.*

(ii) *Considering the function ψ as a map on $\Omega_0 \times (0,\infty)$ we assume that $\psi \in C_b^{3,1}$.*

(iii) *The map ϕ is volume preserving, i.e., we have $\det \nabla_\xi(\phi(\cdot,t)) = 1$.*

(iv) *We assume that $\phi(\cdot,t) \to \phi(\cdot,\infty)$ as $t \to \infty$ in $C^3(\Omega_0)$ for some function $\phi(\cdot,\infty) \in C^3(\overline{\Omega_0})$ and $\partial_t\phi(\cdot,t) \to 0$ as $t \to \infty$ in $C^1(\Omega_0)$.*

(v) *There exists an $R > 0$ such that $\phi(\xi,t) = \xi$ for all $|\xi| > R$.*

Here, the assumption (iv) means that the domain is converging to a fixed domain as $t \to \infty$. This domain is given by $\psi(\Omega_0,\infty)$ and will be denoted by Ω_∞. Furthermore, (v) can be interpreted as that the variation of the domain is only acting locally in space.

To state our main Theorem 2.0.2 we need to introduce the function space $L^p(0,\infty; L^q(\Omega(t)))$. We say, that a function $f \in L^p(0,\infty; L^q(\Omega(t)))$, if the extended function

$$\tilde{f}\colon (x,t) \mapsto \begin{cases} f(x,t), & x \in \Omega(t), \\ 0, & \text{else}, \end{cases}$$

is an element in $L^p(0,\infty; L^q(\mathbb{R}^3))$ and we define

$$(2.3) \qquad \|f\|_{L^p(0,\infty;L^q(\Omega(t)))} := \|\tilde{f}\|_{L^p(0,\infty;L^q(\mathbb{R}^3))}, \quad 1 \le p, q \le \infty.$$

Note that if $1 \le p, q < \infty$ then

$$\tilde{X} := C_0^\infty(0,\infty; C_0^\infty(\Omega(t)))$$
$$:= \{\varphi \in C_0^\infty(0,\infty; C_0^\infty(\mathbb{R}^3)) \mid \operatorname{supp} \varphi(t) \subset \Omega(t) \text{ for all } t \in (0,\infty)\}$$

is a dense subspace of $L^p(0,\infty; L^q(\Omega(t)))$. This can be seen as follows. Choose $f \in L^p(0,\infty; L^q(\Omega(t)))$ and consider the function

$$g := f \circ \psi \in L^p(0,\infty; L^q(\Omega_0)).$$

Then there exists a sequence $(g_n)_n \subset C_0^0(0,\infty; C_0^0(\Omega_0))$ which approximates g and the sequence $(f_n)_n$ defined by $f_n := g_n \circ \psi^{-1}$ approximates f. Furthermore, the functions g_n are continuous, compactly supported in time, and $\operatorname{supp}(g_n(t)) \subset \Omega(t)$ for all $t \in (0,\infty)$. Then, by considering a standard mollifier argument, we can ensure that the approximative sequence can be assumed to be smooth. Due to

$$\inf\{\operatorname{dist}(\operatorname{supp}(g_n(t)), \partial\Omega(t)) \mid t \ge 0\} > 0,$$

which holds true since g_n has compact support in time, we obtain that the smooth approximative sequence can be chosen to be compactly supported in Q.

Hence, an equivalent definition of the space $L^p(0,\infty; L^q(\Omega(t)))$ is provided as the closure of $\tilde{X}$ w.r.t. to the norm of the extended function defined in (2.3). Analogously, we can define the spaces $L^p(0,\infty; L^q_\sigma(\Omega(t)))$ as the closure of $C_0^\infty(0,\infty; C_{0,\sigma}^\infty(\Omega(t)))$ w.r.t. the norm of the extended function and $L^p(0,\infty; \dot{W}^{1,q}(\Omega(t)))$ as the closure of $\tilde{X}$ w.r.t. the norm of $\nabla\tilde{f}$, where $\tilde{f}$ denotes the extended function as before.

Now our main theorem reads as follows.

Theorem 2.0.2. *Let $1 < q < \frac{3}{2}$, $1 < p < \infty$, and let the evolution of $\Omega(t)$ be determined by a function ψ which fulfils Assumption 2.0.1. Then for all $f \in X := \mathrm{L}^p(0, \infty; \mathrm{L}^q_\sigma(\Omega(t)))$ and all $v_0 \in \mathscr{I}^p_{1+A_{\Omega_0}}$ there exists a unique strong solution (v, p) to (2.1) such that $(v(t), p(t)) \in \mathcal{D}(A_{\Omega(t)}) \times \dot{W}^{1,q}(\Omega(t))$ for almost all $t \in (0, \infty)$. Furthermore, there exists a $c > 0$ such that*

$$(2.4) \qquad \|v_t\|_X + \|A_{\Omega(\cdot)} v(\cdot)\|_X \leq c \Big(\|f\|_X + \|v_0\|_{\mathscr{I}^p_{1+A_{\Omega_0}}} \Big).$$

For the definition of the space of initial values $\mathscr{I}^p_{1+A_{\Omega_0}}$ we refer to Section 1.3. Note that $1 + A_{\Omega_0}$ denotes the shifted Stokes operator on Ω_0 and the space $\mathscr{I}^p_{1+A_{\Omega_0}}$ coincides with equivalent norms to the real interpolation space

$$\Big(\mathrm{L}^q_\sigma(\Omega), \mathcal{D}(A_{\Omega_0}) \Big)_{1-\frac{1}{p},p},$$

where $\mathcal{D}(A_{\Omega_0})$ is endowed with the graph norm.

The structure of this chapter is as follows: We will start to transform system (2.1) to a problem in a cylindrical time-space domain in Section 2.1. This will lead to a non-autonomous system and we will present a result to obtain maximal regularity to non-autonomous systems in Section 2.2. The result in this section is due to M. Giga, Y. Giga, and H. Sohr [23]. Fifteen years later, J. Saal [45] proved that the assumptions of this theorem can be alleviated if the family of operators is invertible. This result will be presented in Section 2.3. Finally, we will prove the Main Theorem 2.0.2 in the last Section 2.4.

2.1 Coordinate Transformation

The main idea to deal with many partial differential equations on a non-cylindrical time-space domain as (2.1) is to transform this system to a problem on a cylindrical domain. Therefore, let us introduce the operator

$$(2.5) \qquad \Phi(t) \colon v \mapsto (\Phi(t)v)(\cdot, t) := u(\cdot, t) := (\nabla \phi)^{-1}(\cdot, t) v(\phi(\cdot, t), t).$$

For each $t \in [0, \infty]$, the operator $\Phi(t)$ maps functions defined on the domain $\Omega(t)$ to functions defined on Ω_0. Note that this kind of transformation was studied by several authors before; let us mention in particular [28, 32, 40, 45, 52]. The function $\Phi(t)$ is the main ingredient to transform system (2.1) to a system on a cylindrical time-space domain. One of the most important properties of this transformation is the following regularity result.

Lemma 2.1.1. *Let $\Phi(t)$ be defined as in (2.5) and let the function $\phi(\cdot, t)$ fulfil Assumption 2.0.1. Then the mapping $\Phi(t)$ defines an isomorphism of the following spaces:*

(i) *$\Phi(t)$ is an isomorphism of $L^q(\Omega(t))$ to $L^q(\Omega_0)$ for $1 \leq q \leq \infty$.*

(ii) *Let $1 \leq q \leq \infty$ and $k = 1, 2$. Then the mapping $\Phi(t)$ is a isomorphism of $W^{k,q}(\Omega(t))$ to $W^{k,q}(\Omega_0)$.*

(iii) *For $1 < q < \infty$ the operator $\Phi(t)$ defines a mapping from $W_0^{1,q}(\Omega(t))$ to $W_0^{1,q}(\Omega_0)$. If both spaces are endowed with homogeneous norms, we have that $\Phi(t)$ is an isomorphism provided that $q < 3$.*

(iv) *Finally for $1 \leq q < \infty$, we have that $\Phi(t)$ is also an isomorphism from $L_\sigma^q(\Omega(t))$ to $L_\sigma^q(\Omega_0)$.*

Proof. Note that the inverse of $\Phi(t)$ is given by

$$\Phi(t)^{-1} \colon u \mapsto v = (\nabla \phi^{-1})^{-1} u(\phi^{-1}(\cdot)).$$

Since

$$\nabla_\xi \phi(\xi, t) = (\nabla_x(\phi^{-1}(x, t)))^{-1}$$

by the Inverse Mapping Theorem, where we used the notation $\phi(\xi, t) = x$, we obtain that by assumption $(\nabla_x(\phi^{-1}(x, t)))^{-1}$ is continuous and uniformly bounded w.r.t. to the spatial variable x.

(i) Due to the uniform boundedness of $(\nabla \phi^{-1}(\cdot, t))^{-1}$ we obtain that $\Phi(t)^{-1} \colon L^q(\Omega(t)) \to L^q(\Omega_0)$ is a bounded operator and since $\Phi(t)^{-1}$ is bijective with inverse $\Phi(t)$, we obtain by the Open Mapping Theorem, that the mapping $\Phi(t)$ has the desired regularity.

(ii) That $\Phi(t) \colon W^{k,q}(\Omega(t)) \to W^{k,q}(\Omega_0)$ follows immediately from the chain and product rule. The proof of the continuity can be done as in part (i).

(iii) Note that $C_0^2(\Omega_0)$ or $C_0^2(\Omega(t))$ is a dense subspace of $W_0^{1,q}(\Omega_0)$ or $W_0^{1,q}(\Omega(t))$, respectively. Obviously, $\Phi(t)$ maps a C^2-function onto a C^2-function and hence

$$\Phi(t)(C_0^2(\Omega(t)) \subset C_0^2(\Omega_0).$$

Since the same holds true for the inverse, we obtain

$$\Phi(t)(W_0^{1,q}(\Omega(t))) = W_0^{1,q}(\Omega_0).$$

For the estimate of the inverse norms, note that $\phi - \mathrm{Id}$ has compact support in $\Omega \cap B_R$ and R is chosen as in Assumption 2.0.1, and hence the estimate is due to the Sobolev inequalities.

(iv) As in part (iii), note that $C_{0,\sigma}^2(\Omega(t))$ is a dense subspace of $\mathrm{L}_\sigma^q(\Omega(t))$. Since

$$\mathrm{div}\,_x v = \mathrm{div}\,_\xi \Phi(t)v$$

for all $v \in C_0^2(\Omega(t))$, we obtain that $\Phi(t)$ maps a dense subspace of $\mathrm{L}_\sigma^q(\Omega(t))$ onto a dense subspace of $\mathrm{L}_\sigma^q(\Omega_0)$. □

In Lemma 2.1.1, the regularity statements to the operators $\Phi(t)$ are not *a priori* uniformly in t. But it is obvious and mentioned in the proof, that the norm of the operator $\Phi(t)\colon \mathrm{L}^q(\Omega(t)) \to \mathrm{L}^q(\Omega_0)$ just depends on the L^∞-norm of $\nabla\phi(\cdot, t)$ and its inverse. To show that those norms can be bounded uniformly will be done later.

Using the transformation $\Phi(t)$, we transform the Stokes system (2.1) onto the cylindrical time-space domain $\Omega_0 \times (0, \infty)$. Note that due to

$$\tilde{A}(t) := \Phi(t)P_{\Omega(t)}(-\Delta)\Phi(t)^{-1} = \Phi(t)P_{\Omega(t)}\Phi(t)^{-1}\Phi(t)(-\Delta)\Phi(t)^{-1}$$

we can transform the Helmholtz projection and the Laplacian separately. Let us also remark, that we have some freedom in the choice of the cylindrical time-space domain since for any $t_0 \in (0, \infty)$, by considering the transformation $\Phi_{t_0}(t) := \Phi(t_0)^{-1}\Phi(t)$, we could also transform the Stokes equations onto $\Omega(t_0) \times (0, \infty)$.

For the transformation of the Laplacian we need two preliminary results.

Lemma 2.1.2. *Let* $m \in \mathbb{N}$, $C > 0$, *and let* P *be a polynomial in the variables* $X_1, ..., X_m \in \mathbb{R}$ *such that* $P(0, ..., 0) = 0$. *Then there exists a constant* $c > 0$ *such that*

$$\left| P(X_1, ..., X_m) - P(Y_1, ..., Y_m) \right| \le c \sup_{i=1,...,m} |X_i - Y_i|$$

hold for all $(X_1, ..., X_m), (Y_1, ..., Y_m) \in \mathbb{R}^m$ *such that* $|X_i|, |Y_i| \le C$, $1 \le i \le m$.

Proof. Since P is a C^1-function and the set

$$X = \left\{ (X_1, ..., X_m) \in \mathbb{R}^m, \ |X_i| \le C \right\}$$

is compact, the result follows by the Mean Value Theorem. $\qquad\square$

Lemma 2.1.3. *Let* X *denote a Banach space and* $T \colon X \to \mathcal{L}(X)$ *be such that*
$$\|T(x) - \mathrm{Id}\|_{\mathcal{L}(X)} \le \gamma$$
for all $x \in X$ *and some* $\gamma \in (0, \frac{1}{2})$. *Then* $T(x)$ *is boundedly invertible and*

$$\|T(x)^{-1}\|_{\mathcal{L}(X)} \le 1 + 2\gamma.$$

Proof. Let $\tilde{T} := \mathrm{Id} - T$. Then due to the Neumann series we get that

$$T(x)^{-1} = \left(\mathrm{Id}_X - \tilde{T}(x) \right)^{-1}$$
$$= \sum_{k=0}^{\infty} \left(\tilde{T}(x) \right)^k,$$

and hence
$$\|T(x)^{-1}\|_{\mathcal{L}(X)} \le 1 + 2\|\tilde{T}(x)\|_{\mathcal{L}(X)} \le 1 + 2\gamma. \qquad\square$$

Note that in the application afterwards the Banach space in the last lemma will be equal $\mathbb{R}^{3\times 3}$. In that situation we have a concrete representation of the inverse by the Theorem of Cayley-Hamilton and the proof could be simplified. For sake of completeness we have given a proof in the general situation.

Corollary 2.1.4. *Let X be a Banach space and for a domain $\Omega \subset \mathbb{R}^n$ let $\Upsilon \colon \Omega \to \mathcal{L}(X)$ be a continuous map such that $\Upsilon(x)$ is invertible for all $x \in \Omega$. Then the map $x \mapsto \Upsilon(x)^{-1}$ is continuous.*

Proof. Let $x \in \Omega$ be arbitrary and consider the map

$$\kappa_x \colon \xi \mapsto \Upsilon(\xi)\Upsilon(x)^{-1}.$$

Then the map κ_x is continuous and $\kappa_x(x) = \mathrm{Id}$, hence there exists a neighbourhood U of x such that

$$\|\kappa_x(\xi) - \mathrm{Id}\|_{\mathcal{L}(X)} \le \frac{1}{4}$$

for all $\xi \in U$. Thus, Lemma 2.1.3 implies that $\Upsilon(x)\Upsilon(\xi)^{-1}$ is uniformly bounded for $\xi \in U$. In particular, $\Upsilon(\xi)^{-1}$ is uniformly bounded. Then using

$$\Upsilon(\xi)^{-1} - \Upsilon(x)^{-1} = \Upsilon(\xi)^{-1}\Big(\Upsilon(x) - \Upsilon(\xi)\Big)\Upsilon(x)^{-1}$$

we obtain the continuity of the map $\xi \mapsto \Upsilon(\xi)^{-1}$. $\qquad\square$

Corollary 2.1.5. *Let ϕ fulfil Assumption 2.0.1. Then $\nabla_x \phi^{-1}$ is uniformly bounded.*

Proof. The last corollary implies that

$$(x,t) \mapsto \nabla_x \phi^{-1}(x,t) = \Big(\nabla_\xi \phi(\phi^{-1}(x,t),t)\Big)^{-1}$$

is continuous. Furthermore, Assumption 2.0.1 (v), implies that

$$\nabla_x \phi^{-1}(x,t) = \mathrm{Id}$$

if $|x|$ is sufficiently large and (iv) yields that $\nabla\phi^{-1}(x,t)$ converges if $t \to \infty$. Hence by considering the one-point compactification in time, we obtain the result. $\qquad\square$

Corollary 2.1.6. *The norm of the isomorphism $\Phi(t)$ in the spaces listed in Lemma 2.1.1 (i) is bounded uniformly in t.*

Proof. As discussed after the proof of Lemma 2.1.1, the norm of $\Phi(t)$ and $\Phi(t)^{-1}$ can be bounded in terms of $\|\nabla_\xi \phi(\cdot,t)\|_\infty$ and $\|\nabla_x \phi^{-1}(\cdot,t)\|_\infty$ and these norms are bounded uniformly in t. $\qquad\square$

Let us now start to transform the Laplacian.

Lemma 2.1.7. *Let ϕ and ψ fulfil Assumption 2.0.1 with $\|\phi - \mathrm{Id}\|_{C_b^{3,1}} \leq \frac{1}{2}$. Let $\Phi(\cdot)$ be defined as in (2.5). Then the following holds:*

(i) *We have that*

$$\Phi(t)\Delta\Phi(t)^{-1} = \Delta + \sum_{|\alpha|\leq 2} a_\alpha(\xi, t)\partial^\alpha,$$

for some matrix-valued coefficient functions a_α. It holds that

$$a_\alpha(\cdot, 0) = 0.$$

Furthermore, there exists a $c > 0$ such that the estimate

$$\|a_\alpha(\cdot, t) - a_\alpha(\cdot, \tau)\|_{\mathrm{L}^\infty(\Omega_0)} \leq c|\phi(t) - \phi(\tau)|_{C_b^{3,1}}$$

for all $|\alpha| \leq 2$ and $t, \tau \in [0, \infty)$ holds. Here, the constant c can be chosen independently of t and τ. In particular, we have

$$\|a_\alpha(\cdot, t)\|_{\mathrm{L}^\infty(\Omega_0)} \leq c\|\phi - \mathrm{Id}\|_{C_b^{3,1}}.$$

Furthermore, we have that

$$supp(a_\alpha) \subset \overline{B_R}.$$

(ii) *It holds that*

$$\Phi(t)\partial_t\Phi(t)^{-1} = \partial_t + \sum_{|\beta|\leq 1} b_\beta(\xi, t)\partial^\beta$$

for some matrix valued, compactly supported coefficient functions b_β. Furthermore, there exists a $c > 0$ which does not depend on t and τ such that

$$\|b_\beta(\cdot, t) - b_\beta(\cdot, \tau)\|_{\mathrm{L}^\infty(\Omega_0)} \leq c|\phi(t) - \phi(\tau)|_{C_b^{3,1}}$$

for all $|\beta| \leq 1$ and $t, \tau \in [0, \infty)$, and we have

$$\lim_{t\to\infty} b_\beta(\cdot, t) = 0.$$

In particular, we obtain

$$\|b_\beta(\cdot, t)\|_{\mathrm{L}^\infty(\Omega_0)} \leq c\|\phi - \mathrm{Id}\|_{C_b^{3,1}}.$$

Proof. An elementary calculation yields that

(2.6)

$$(\Phi(t)\Delta_x\Phi(t)^{-1}u)(\xi,t) = \sum_{i,j,k,l,m=1}^{3} \left(\partial_{x_k}\phi^{-1}\right)\left(\partial_{x_j}\phi^{-1}\right)_i \left(\partial_{x_j}\phi^{-1}\right)_l (\phi(\xi,t),t)$$

$$\cdot \left[\left(\partial_{\xi_l}\partial_{\xi_i}\partial_{\xi_m}\phi_k\right)u_m + \left(\partial_{\xi_i}\partial_{\xi_m}\phi_k\right)\partial_{\xi_l}u_m\right.$$

$$\left. + \left(\partial_{\xi_l}\partial_{\xi_m}\phi_k\right)\partial_{\xi_i}u_m + \left(\partial_{\xi_m}\phi_k\right)\partial_{\xi_l}\partial_{\xi_i}u_m\right].$$

The exact representation of the transformed Laplacian will not be important in the following. Let us just point out that the transformed Laplacian can be rewritten as a polynomial in the variables $\nabla^k\phi$, $1 \leq k \leq 3$, and $\nabla\phi^{-1}$.

Furthermore, since $\Phi(0) = \text{Id}$ we obtain that

$$\Phi(0)\Delta_x\Phi(0)^{-1} = \Delta$$

Hence let us define the coefficients a_α such that

$$\Phi(t)\Delta\Phi(t)^{-1} - \Delta = \sum_{|\alpha|\leq 2} a_\alpha(x,t)\partial^\alpha$$

holds true. Then, obviously, $a_\alpha(\cdot,0) = 0$. Furthermore, we have that outside the ball $\overline{B}_R$ the coefficient functions vanish. This can be seen easily since the right-hand side in the representation formula for $\Phi(t)\Delta_x\Phi(t)^{-1}$ vanishes if either $m = k$ or $i = j = l$ is not fulfilled and in that case we see that the right-hand side is equal to Δu.

It remains to prove the estimate for a_α. Therefore, note that a_α can be rewritten as a polynomial with coefficients in $\nabla^k\phi$, $0 \leq k \leq 2$ and $\nabla\phi^{-1}$. Thus by assumption we have that $\|\partial^k\phi\|_\infty \leq \gamma := \|\phi - \text{Id}\|_{C_b^{3,1}}$. Due to Lemma 2.1.3 we have that all coefficients are bounded and

$$\nabla\phi^{-1}(x,t) - \nabla\phi^{-1}(y,\tau) = \nabla\phi^{-1}(x,t)\left(\nabla\phi(\xi,\tau) - \nabla\phi(\xi,t)\right)\nabla\phi^{-1}(y,\tau)$$

with $x = \phi(\xi,\tau)$ and $y = \phi(\xi,t)$. Hence Lemma 2.1.2 implies the estimate

$$\|a_\alpha(\cdot,\tau) - a_\alpha(\cdot,t)\|_\infty \leq c|\phi(\tau) - \phi(t)|_{C_b^{3,1}}.$$

Thus, we also get

$$\|a_\alpha(\cdot,t)\|_\infty = \|a_\alpha(\cdot,t) - a_\alpha(\cdot,0)\|_\infty \leq c|\phi(t) - \mathrm{Id}|_{C_b^{3,1}}.$$

The idea to prove the properties of b_β is analogous. Note that we have

$$\Phi(t)\partial_t\Phi(t)^{-1}u$$

$$= \partial_t u + (\nabla\phi)^{-1}\left(\sum_{i,j}(\partial_t\phi)_j^{-1}[\partial_{\xi_i}\partial_{\xi_j}\phi\, u_i + \partial_{\xi_i}\phi\, \partial_{\xi_j}u_i] + \sum_i(\partial_{\xi_i}\partial_t\phi)u_i\right).$$

The term in the bracket vanishes outside $\overline{B}_R$ since the time derivative and second derivatives of ϕ have a support in $\overline{B}_R$. The proof of the estimate can be done analogously to the proof of the estimate to a_α. $\qquad\square$

In the same manner as in the last Lemma 2.1.7 it is also possible to transform the Helmholtz projection $P_{\Omega(t)}$ on $\Omega(t)$ into a functions defined on function on Ω_0. Therefore, let us define

$$(2.7) \qquad P(t) := \Phi(t)P_{\Omega(t)}\Phi(t)^{-1}.$$

Since $\Phi(t)$ is an isomorphism of $\mathrm{L}_\sigma^q(\Omega(t))$ to $\mathrm{L}_\sigma^q(\Omega_0)$, we see that $P(t)$ is a projection on $\mathrm{L}_\sigma^q(\Omega_0)$. But in general there is no reason to assume that the kernel of $P(t)$ coincides with the kernel of P_{Ω_0}, hence $P(t)$ might differ from the Helmholtz projection on Ω_0. Nevertheless, the following result to the evolution of $P(\cdot)$ was proved by J. Saal, [45, Lemma 3.1].

Lemma 2.1.8. *There exists a $c > 0$ such that for all $t, s \geq 0$ the estimate*

$$(2.8) \qquad \|P(t) - P(s)\|_{\mathcal{L}(\mathrm{L}^q(\Omega_0))} \leq c\|(\nabla\phi)(t)^{-1} - (\nabla\phi)(s)^{-1}\|_\infty$$

holds true. In particular, $\left(\|P(t)\|_{\mathcal{L}(\mathrm{L}^q(\Omega_0))}\right)_{0\leq t<\infty}$ is bounded.

Let us now define the transformed operators.

Definition 2.1.9. Let ψ and ϕ fulfil Assumption 2.0.1, let a_α, b_β be as in Lemma 2.1.7, and $P(t)$ be defined as in (2.7). Furthermore, let $1 < q < \frac{3}{2}$. Then we define the operators $A(t)$ by

$$A(t)\colon \mathcal{D}(A) \to \mathrm{L}_\sigma^q(\Omega_0),$$

$$u \mapsto P(t)\left(-\Delta - \sum_{|\alpha|\leq 2} a_\alpha(\cdot,t)\partial^\alpha + \sum_{|\beta|\leq 1} b_\beta(\cdot,t)\partial^\beta\right)u.$$

Here, $\mathcal{D}(A)$ denotes the domain of the Stokes operator on Ω_0, hence

$$\mathcal{D}(A) = W^{2,q}(\Omega_0) \cap W_0^{1,q}(\Omega_0) \cap L_\sigma^q(\Omega_0).$$

Furthermore, let us denote by $\tilde{A}(t)$ the transformed Stokes operator

$$\tilde{A}(t) \colon \mathcal{D}(A) \to L_\sigma^q(\Omega_0),$$
$$u \mapsto P(t)\Big(-\Delta - \sum_{|\alpha| \leq 2} a_\alpha(\cdot, t)\partial^\alpha\Big)u,$$

and let
$$B(t) := A(t) - \tilde{A}(t).$$

Note that since $\Phi(t)$ is an isomorphism of the domain of the Stokes operator, we obtain that the operator $\tilde{A}(t)$ coincides with the transformed Stokes operator, hence

$$\tilde{A}(t) = \Phi(t)A_{\Omega(t)}\Phi(t)^{-1}.$$

Using that we rewrite Theorem 2.0.2 as follows.

Theorem 2.1.10. *Let $1 < q < \frac{3}{2}$, $1 < p < \infty$ and let the evolution of $\Omega(t)$ be determined by a function ψ which fulfils Assumption 2.0.1. Furthermore, let $A(\cdot)$ be defined as in Definition 2.1.9. Then for all $f \in X := L^p(0, \infty; L_\sigma^q(\Omega_0))$ and all $u_0 \in \mathscr{I}_{1+\tilde{A}(0)}^p$ there exists a strong solution u to*

$$(2.9) \qquad\qquad u_t + A(t)u = f, \; u(0) = u_0.$$

Furthermore, there exists a constant $c > 0$ such that the estimate

$$(2.10) \qquad \|u_t\|_X + \|A(\cdot)u\|_X \leq c\Big(\|f\|_X + \|u_0\|_{\mathscr{I}_{1+\tilde{A}(0)}^p}\Big)$$

holds true.

Throughout the rest of this chapter, we will focus on the proof of Theorem 2.1.10. Note that a function v defined on the non-cylindrical time-space domain Q is a strong solution to (2.1) if and only if $u := \Phi(\cdot)v$ is a strong solution to the system (2.9). Since u is defined on a cylindrical time-space domain, it seems to be easier to solve the non-autonomous system (2.9). Unfortunately, the maximal regularity estimate (2.4) for the solution v is equivalent to the estimate

$$(2.11) \qquad \|(\partial_t + B(\cdot))u\|_{q,p} + \|\tilde{A}(\cdot)u\|_{q,p} \le c\big(\|\Phi(\cdot)f\|_{q,p} + \|v_0\|\big)$$

for the solution u due to the isomorphism property of Φ. But it is not obvious if the estimate (2.11) implies (2.10), and hence the equivalence of Theorem 2.0.2 and Theorem 2.1.10 has to be proven carefully. This will be done at the end of this chapter.

In the subsequent Section 2.2 we will present a strategy to deal with such kind of problems.

Before we start to introduce the technique to prove a maximal regularity result for non-autonomous systems, we need to introduce a second coordinate transform. Note that it is possible to change the mapping ϕ in the definition of $\Phi(\cdot)$ in (2.5), to receive a coordinate transform to another reference domain. Therefore, let $t_0 \in (0, \infty]$ and let us define

$$\phi_{t_0}(t) := \phi(\cdot, t_0)^{-1} \circ \phi(\cdot, t).$$

Obviously, $\phi_{t_0}(\cdot)$ defines a family of isomorphisms with the same regularity properties as $\phi(\cdot)$, and we can introduce the transformation map

$$(2.12) \qquad \Phi_{t_0}(t) \colon v \mapsto ((\nabla \phi_{t_0})^{-1}(\cdot, t))v(\phi_{t_0}(\cdot, t)),$$

defined similarly to $\Phi(t)$. Correspondingly, we obtain the same regularity results.

Lemma 2.1.11. *Let Φ_{t_0} be defined as in (2.12) and let $\phi(\cdot, t)$ and $\phi_{t_0}(\cdot, t)$ fulfil Assumption 2.0.1.*

(i) *The regularity results for $\Phi(\cdot)$ in Lemma 2.1.1 remain true if we replace Φ by Φ_{t_0}.*

(ii) *Let us define $P_{t_0}(t) := \Phi_{t_0}(t)P_{\Omega(t)}\Phi_{t_0}(t)^{-1}$. Then Lemma 2.1.8 remains true if we replace $P(t)$ by $P_{t_0}(t)$.*

(iii) *Let us introduce the operators $A_{t_0}(\cdot)$, $\tilde{A}_{t_0}(\cdot)$ and $B_{t_0}(\cdot)$ defined analogously to A, $\tilde{A}$ and B in Definition 2.1.9 by replacing $\Phi(\cdot)$ with $\Phi_{t_0}(\cdot)$. Furthermore, let $a_\alpha^{t_0}$ and $b_\beta^{t_0}$ be defined analogously to Lemma 2.1.7. Then $a_\alpha^{t_0}$ and $b_\beta^{t_0}$ have the same properties as a_α, b_β, which are listed in Lemma 2.1.7.*

Proof. Since in the proof of the corresponding results in the Lemmas 2.1.1, 2.1.7, and 2.1.8 we just used the regularity properties of ϕ, and since ϕ_{t_0} has the same regularity properties, the proofs carry over one-to-one. □

Before we end the section about the coordinate transformation let us introduce the operator $\tilde{A}_\infty$. Since the map $\phi(\cdot, t)$ converges as $t \to \infty$ in C^3, we obtain that the coefficient $a_\alpha(\cdot, t)$ converge to some coefficient function $a_\alpha(\cdot, \infty)$ in $L^\infty(\Omega_0)$. Furthermore, the estimate in Lemma 2.1.8 implies that

$$P(\infty) := \lim_{t\to\infty} P(t) \in \mathcal{L}(L^q(\Omega_0))$$

exists. Let us prove that the transformed Stokes operator of the domain Ω_∞ coincides with the limit of the operator $\tilde{A}(t)$ as $t \to \infty$.

Lemma 2.1.12. *Let ψ and ϕ fulfil Assumption 2.0.1 and let $1 < q < \infty$. Then*

$$\Phi(\infty) A_{\Omega_\infty} \Phi(\infty)^{-1} = P(\infty)\Big(\Delta + \sum_\alpha a_\alpha(\cdot, \infty)\partial^\alpha\Big),$$

where A_{Ω_∞} denotes the Stokes operator on Ω_∞.

Proof. Let us start to prove that $\lim_{t\to\infty} P(t)$ exists. Due to (2.8), we see that $(P(n))_{n\in\mathbb{N}}$ is a Cauchy sequence in $\mathcal{L}(L^q(\Omega_0))$ and hence there exists an operator $P(\infty)$ as the limit of $(P(n))_n$. Furthermore, (2.8) also implies that

$$P(\infty) = \lim_{t\to\infty} P(t) \in \mathcal{L}(L^q(\Omega_0)).$$

Since the proof of Lemma 2.1.8 is also valid for $s = \infty$, we obtain that

$$P(\infty) = \Phi(\infty) P_{\Omega_\infty} \Phi(\infty)^{-1},$$

see [45, Theorem 3.2]. Furthermore, the convergence of $\phi(\cdot, t) \to \phi(\cdot, \infty)$ in $C^3(\Omega_0)$ implies that

$$\lim_{t\to\infty} a_\alpha(\cdot, t) = a_\alpha(\cdot, \infty) \in L^\infty(\Omega_0)$$

for all α. Thus, we have

$$\Phi(\infty)\Delta\Phi(\infty)^{-1} = \Big(\Delta + \sum_\alpha a_\alpha(\cdot, \infty)\partial^\alpha\Big).$$

This proves the assertion. □

2.2 Non-Autonomous Systems

The most important result to prove Theorem 2.1.10 is a result to obtain maximal regularity for non-autonomous systems by M. Giga, Y. Giga, and H. Sohr [23, Theorem]. Due to the importance of their result and in particular since we changed the assumptions slightly, we will present a proof for this result. Since our main result deals with global in time strong solutions, we focused in the last sections to the case if the time interval is $[0, \infty)$. Since the next results will also be valid if the time interval is finite, we will present the result for both cases.

Definition 2.2.1. Let $0 < T \leq \infty$ and let X be a Banach space. Let $A(t)$ for $t \in [0, T)$ denote a closed operator. Then we say that the non-autonomous system

$$(2.13) \qquad u_t + A(\cdot)u = f, \; u(0) = u_0$$

has maximal L^p-regularity, if for all $f \in L^p(0, T; X)$ and all $u_0 \in \mathscr{I}^p_{A(0)}$ there exists a unique strong solution to (2.13) and there exists a constant $c > 0$ such that

$$(2.14) \qquad \|u_t\|_{X,p} + \|A(\cdot)u\|_{X,p} \leq c\Big(\|f\|_{X,p} + \|u_0\|_{\mathscr{I}^p_{A(0)}}\Big).$$

Theorem 2.2.2. *Let* $0 < T \leq \infty$ *and let* X *be a Banach space. Let* $A(t)$ *for* $t \in [0, T)$ *denote a closed operator and let the following assumptions be fulfilled:*

(i) *For every* $t \in [0, T)$ *the operator* $A(t)$ *is densely defined and its range* $\mathrm{Im}(A(t))$ *is dense. Furthermore,* $\mathcal{D}(A(t))$ *does not depend on* t *and* $A(t)$ *is injective for all* $t \in [0, T)$.

(ii) *For every* $\lambda > 0$ *we have* $(\lambda + A(t))^{-1} \in \mathcal{L}(X)$ *and there are constants* $M(t) > 0$ *such that* $\|\lambda(\lambda + A(t))^{-1}\|_{\mathcal{L}(X)} \leq M(t)$.

(iii) *There exists a* $c_1 > 0$ *such that* $\|A(t)x\|_X \leq c_1\|A(\tau)x\|_X$ *for every* $0 \leq \tau < t < T$ *and all* $x \in \mathcal{D}(A(\tau))$. *Thus, the operator* $A(t)A(\tau)^{-1}$ *extends to a bounded operator in* $\mathcal{L}(X)$.

Furthermore, the map

$$[0, T) \to \mathcal{L}(X), \quad t \mapsto A(t)A(0)^{-1}$$

is continuous.

(iv) *There exists a $c_2 > 0$ such that $A(t) \in \mathcal{MR}_p(X, c_2)$ holds for all $t \in [0, T)$.*

(v) *There exists an $\varepsilon \in (0, c_2^{-1})$ and finitely many $0 = T_0 < ... < T_n = T$ such that*

$$\|(A(t) - A(T_i))A(T_i)^{-1}x\|_X < \varepsilon$$

for all $i = 0, ..., n - 1$, $t \in [T_i, T_{i+1})$ and $x \in X$ such that $\|x\|_X = 1$.

Then the non-autonomous system

(2.15)
$$u_t + A(\cdot)u = f, \ u(0) = u_0$$

has maximal L^p-regularity.

The space of initial values in the theorem above was defined in (1.9). If $T = \infty$, due to Lemma 1.3.3, this space coincides with the real interpolation space

$$\mathscr{I}_{A(0)}^p = \left(X, \mathcal{D}(A(0))\right)_{1-\frac{1}{p}, p}$$

and the domain of $A(0)$ is endowed with the homogeneous norm.

Remark 2.2.3. There are some difference in our formulation of Theorem 2.2.2 to the Theorem in [23].

(i) In [23] it is assumed that the Banach space X is of class $\mathcal{HT}$, that the operators $A(t)$ have bounded imaginary powers, and the power angles can be bounded uniformly by a constant less that $\frac{\pi}{2}$ and furthermore, the constant in the corresponding estimate is bounded uniformly. Those properties are just used to prove assumption (iv) in our version.

(ii) Note that the assumptions a) and b) in [23, Theorem] on $A(t)$, which are almost the same as (i) and (ii) in our version, are just equivalent to that $A(t)$ generates an analytic semigroup.

(iii) Instead of (v), in [23] it is assumed that the map

$$\{(t, \tau) \in [0, T)^2 \mid t \geq \tau\} \to \mathcal{L}(X)$$
$$(t, \tau) \mapsto A(t)A(\tau)^{-1}$$

is continuous and if $T = \infty$, $A(t)A(\tau)^{-1} \to$ Id if $t \geq \tau \to \infty$. Obviously, the assumptions in [23] imply (v), but the converse does not hold. For example, we could also treat the case if $t \mapsto A(t)$ is periodic with a small deviation.

Let us start with the proof of Theorem 2.2.2.

Proof of Theorem 2.2.2. Due to (v) there exist finitely many points in time $0 = T_0 < T_1 < ... < T_N = T$ such that

$$\|(A(\tau) - A(T_i))A(T_i)^{-1}\|_{\mathcal{L}(X)} < \varepsilon$$

for all $i = 0, ..., N$ and all $\tau \in [T_i, T_{i+1})$.

Let us assume first that there exists a solution

$$u \in W^{1,p}_{\text{loc}}([0,T); X) \cap L^p_{\text{loc}}([0,T); \mathcal{D}(A))$$

to (2.15). Hence it remains to prove the estimate (2.14). Let u_1 denote a solution to

$$\dot{u}_1 + A(0)u_1 = 0, \ u_1(0) = u_0$$

in $[0, T_1)$ and let $u_2 = u - u_1$. Using (iii) we have

$$\|\dot{u}_1\|_{L^p(0,T_1;X)} + \|A(\cdot)u_1\|_{L^p(0,T_1;X)} \leq c_1\left(\|\dot{u}_1\|_{L^p(0,T_1;X)} + \|A(0)u_1\|_{L^p(0,T_1;X)}\right)$$

$$\leq c_1 c_2 \|u_0\|_{\mathscr{I}^p_{A(0)}}.$$

Furthermore, we obtain

$$\|\dot{u}_2 + A(\cdot)u_2\|_{L^p(0,T_1;X)}$$
$$\geq \|\dot{u}_2 + A(0)u_2\|_{L^p(0,T_1;X)} - \|(A(\cdot) - A(0))u_2\|_{L^p(0,T_1;X)}$$
$$\geq \frac{1}{c_2}\left(\|\dot{u}_2\|_{L^p(0,T_1;X)} + \|A(0)u_2\|_{L^p(0,T_1;X)}\right) - \varepsilon\|A(0)u_2\|_{L^p(0,T_1;X)}$$
$$\geq \frac{1}{c_2}\|\dot{u}_2\|_{L^p(0,T_1;X)}$$

as well as

$$\|A(\cdot)u_2\|_{L^p(0,T_1;X)} \leq \|\dot{u}_2 + A(\cdot)u_2\|_{L^p(0,T_1;X)} + \|\dot{u}_2\|_{L^p(0,T_1;X)}$$
$$\leq (1 + c_2)\|\dot{u}_2 + A(\cdot)u_2\|_{L^p(0,T_1;X)}.$$

Hence we can combine these estimates, and this leads to the estimate

(2.16)
$$\|\dot{u}\|_{L^p(0,T_1;X)} + \|A(\cdot)u\|_{L^p(0,T_1;X)}$$
$$\leq \|\dot{u}_1\|_{L^p(0,T_1;X)} + \|A(\cdot)u_1\|_{L^p(0,T_1;X)} + \|\dot{u}_2\|_{L^p(0,T_1;X)} + \|A(\cdot)u_2\|_{L^p(0,T_1;X)}$$
$$\leq C\Big(\|u_0\|_{\mathscr{I}^p_{A(0)}} + \|\dot{u}_2 + A(\cdot)u_2\|_{L^p(0,T_1;X)}\Big)$$
$$\leq C\Big(\|u_0\|_{\mathscr{I}^p_{A(0)}} + \|\dot{u} + A(\cdot)u\|_{L^p(0,T_1;X)} + \|\dot{u}_1 + A(\cdot)u_1\|_{L^p(0,T_1;X)}\Big)$$
$$\leq C\Big(\|u_0\|_{\mathscr{I}^p_{A(0)}} + \|f\|_{L^p(0,T_1;X)}\Big),$$

where this constant C depends only on c_1 and c_2. Iterating the same arguments in the time intervals $[T_i, T_{i+1})$ we obtain that

$$\|\dot{u}\|_{L^p(0,T;X)} + \|A(\cdot)u\|_{L^p(0,T;X)} \leq C\Big(\|f\|_{L^p(0,T;X)} + \sum_{i=0}^{N-1}\|u(T_i)\|_{\mathscr{I}^p_{A(T_i)}}\Big).$$

Hence it is sufficient to prove that we can estimate the term

$$\sum_{i=1}^{N-1}\|u(T_i)\|_{\mathscr{I}^p_{A(T_i)}}$$

on the right-hand side with the other terms on the right-hand side. We will argue by contradiction. Let us assume that there exists a sequence $(u_n)_n$ such that

$$\|u_n(0)\|_{\mathscr{I}^p_{A(0)}} + \|\dot{u}_n + A(\cdot)u_n\|_{L^p(0,T;X)} \leq \frac{1}{n}$$

and

$$\|\dot{u}_n\|_{L^p(0,T;X)} + \|A(\cdot)u_n\|_{L^p(0,T;X)} = 1.$$

Let us consider the sequence of functions

$$\bar{u}_n \colon t \mapsto \begin{cases} u_n(T_1 - t), & t \leq T_1, \\ e^{-(t-T_1)A(0)}u_n(0), & t \geq T_1. \end{cases}$$

Then Lemma 1.3.3, Remark 1.3.4, and (2.16) imply that

$$\|u_n(T_1)\|_{\mathscr{I}^p_{A(T_1)}} \leq c\Big(\|\dot{\bar{u}}_n\|_{L^p(0,T;X)} + \|A(T_1)\bar{u}_n\|_{L^p(0,T;X)}\Big)$$
$$\leq c\Big(\|\dot{u}_n\|_{L^p(0,T_1;X)} + \|A(\cdot)u_n\|_{L^p(0,T_1;X)} + \|u_n(0)\|_{\mathscr{I}^p_{A(0)}}\Big)$$
$$\leq c\Big(\|f\|_{L^p(0,T_1;X)} + \|u_n(0)\|_{\mathscr{I}^p_{A(0)}}\Big),$$

and thus we have proved the convergence

$$\lim_{n \to \infty} \|u_n(T_1)\|_{\mathscr{I}^p_{A(T_1)}} = 0.$$

Repeating that argument for $\|u_n(T_2)\|, \dots, \|u_n(T_{N-1})\|$ we obtain a contradiction. Thus, the estimate (2.14) is proved.

So far, we proved that if there exists a solution

$$u \in \mathrm{L}^p_{\mathrm{loc}}([0, T); \mathcal{D}(A)) \cap W^{1,p}_{\mathrm{loc}}([0, T); X),$$

then this solution fulfils the maximal regularity estimate (2.14). Thus, we have to prove the existence of a solution to (2.15) in

$$\mathrm{L}^p_{\mathrm{loc}}([0, T); \mathcal{D}(A)) \cap W^{1,p}_{\mathrm{loc}}([0, T); X).$$

Let T_1 and ε as before. Then since $A(0) \in \mathcal{MR}$, we obtain that the operator

$$L_{T_1} := \partial_t + A(0) \colon \dot{W}^{1,p}(0, T_1; X) \cap \mathrm{L}^p(0, T_1; \mathcal{D}(A)) \to \mathrm{L}^p(0, T_1; X),$$
$$u \mapsto \partial_t u + A(0)u,$$

is continuous and continuously invertible. The norm of the inverse is bounded by c_2.

Assumption (v) implies that

$$\|L_{T_1} - (\partial_t + A(\cdot))\|_{\mathcal{L}(\dot{W}^{1,p}(0,T_1;X) \cap \mathrm{L}^p(0,T_1;\mathcal{D}(A)), \mathrm{L}^p(0,T_1;X))}$$
$$= \|A(0) - A(\cdot)\|_{\mathcal{L}(\dot{W}^{1,p}(0,T_1;X) \cap \mathrm{L}^p(0,T_1;\mathcal{D}(A)), \mathrm{L}^p(0,T_1;X))}$$
$$\leq \varepsilon$$
$$< (\|L_{T_1}^{-1}\|_{\mathcal{L}(\mathrm{L}^p(0,T_1;X), \dot{W}^{1,p}(0,T_1;X) \cap \mathrm{L}^p(0,T_1;\mathcal{D}(A)))})^{-1}.$$

Hence by a Neumann series argument we obtain the solvability of (2.15) in the time interval $[0, T_1)$. Repeating that argument in $[T_1, T_2)$ we obtain the solvability of (2.15) in the interval $[0, T_2)$. Inductively, we obtain the solvability of (2.15). □

2.3 The Case of an Invertible Operator

In concrete applications, the main difficulty to apply Theorem 2.2.2 is to prove the uniformity of the constant in the maximal regularity estimate. In 2006 J. Saal proved that one can relinquish the uniformity, if the operators $A(t)$ are invertible. To be more precise, he proved the following, see [45, Theorem 1.4]:

Theorem 2.3.1. *Let X be a Banach space, $1 < p < \infty$, $T \in (0, \infty]$, and $(A(t))_t$ be a family of boundedly invertible operators in X satisfying the following:*

(i) $\mathcal{D}(A(t)) = \mathcal{D}(A(0))$, $t \in (0, T]$.

(ii) *The function $A(\cdot)\colon [0, T) \to \mathcal{L}(\mathcal{D}(A(0)), X)$ is continuous, where $\mathcal{D}(A(0))$ is endowed with the graph norm.*

(iii) $A(t) \to A(T)$ *as $t \to T$ in $\mathcal{L}(\mathcal{D}(A(0)), X)$.*

(iv) *For each $t \in [0, T]$ we have $A(t) \in \mathcal{MR}(X)$.*

Then $A(\cdot)$ has maximal L^p-regularity.

Note that there are two main differences to the assumptions in Theorem 2.2.2. The main one is that in Assumption (iv) there is no uniformity of the constant in the maximal L^p-regularity estimate assumed. In addition, also the continuity of the map $(t, \tau) \mapsto A(t)A(\tau)^{-1}$ as in Theorem 2.2.2 (iii), is not assumed. It is proved that it suffices to assume the continuity in Theorem 2.3.1 (iii).

The idea of the proof is the following. Since the map

$$t \mapsto \|A(t)\|_{\mathcal{L}(\mathcal{D}(A(0)), X)}$$

in continuous and converges as $t \to \infty$, we obtain that this function attains its maxima and minima. Since the continuity of $t \mapsto A(t)$ implies the continuity of $t \mapsto A(t)^{-1}$, we obtain that also $\|A(t)^{-1}\|_{\mathcal{L}(X, \mathcal{D}(A(0)))}$ attains its minima and maxima. Thus, the map $\tau \mapsto L(\tau) := \partial_t + A(\tau)$, *i.e.*, $L(\tau)$ denotes the autonomous differential operator, defines a continuous operator on $[0, T]$ to $\mathcal{L}(\mathbb{E}, \mathrm{L}^p(0, T; X))$ with

$$\mathbb{E} := W^{1,p}(0, T; X) \cap \mathrm{L}^p(0, T; \mathcal{D}(A(0))))$$

and we can argue as above to obtain that $\tau \mapsto \|L(\tau)\|_{\mathcal{L}(\mathbb{E}, \mathrm{L}^p(0,T;X))}$ attains its maxima and minima and the same holds true for its inverse. Hence this proves the uniformity of the maximal L^p-regularity estimate and hence one can apply [23, Theorem].

This result can be applied to a family of transformed Stokes operators if the ground space is a bounded domain or if the ground space is an exterior domain and one adds a shift and hence one is restricted to a result for finite times, see also Lemma 1.3.2.

So far, the results in [45] allow us to treat the non-autonomous system

$$u_t + P(t) \sum_{|\beta| \leq 1} b_\beta(\xi, t) \partial^\beta u - P(t) \Delta u - P(t) \sum_{|\alpha| \leq 2} a_\alpha(\xi, t) \partial^\alpha u = f,$$

$$u(0) = u_0$$

in a finite time interval $[0, T]$. Here we used the notation introduced in Section 2.1. Then, as a second step, by a perturbation result for non-autonomous systems also the missing term $\sum_\beta P(t) b_\beta \partial^\beta$ can be added to the system. Finally, J. Saal proved the following Theorem, see [45, Theorem 1.2]:

Theorem 2.3.2. *Let $1 < p, q < \infty$ and $T \in (0, \infty)$. Furthermore, assume that the evolution of the domains $\Omega(t)$ are described by a mapping ψ which fulfil Assumption 2.0.1 (i)-(iv). Then the system (2.9) has maximal L^p-regularity in $\mathrm{L}_\sigma^q(\Omega_0)$, i.e., for all $f \in \mathrm{L}^p(0, T; \mathrm{L}_\sigma^q(\Omega_0))$ and all initial values $u_0 \in \mathscr{I}_A^p$ there exists a solution u to (2.9) such that the estimate (2.10) holds true with $X = \mathrm{L}^p(0, T; \mathrm{L}_\sigma^q(\Omega_0))$.*

The same kind of Theorem also holds true if the domain is a perturbed half space or if the domain is bounded. In the last case also $T = \infty$ can be included by the method of J. Saal since the Stokes operator on a bounded domain is invertible. Our approach to obtain a result global in time differs mostly in one points. Since we were considering a system with non-invertible operators we cannot copy his elegant argumentation with a stability result for homeomorphisms to obtain the uniformity in the maximal regularity estimate.

The following Lemma 2.3.3 will be crucial to combine J. Saal's result in a finite time interval $[0, T_0)$ for some $T_0 \in (0, \infty)$ with a perturbation argument in $[T_0, \infty)$.

Lemma 2.3.3. *Let* $1 < q < \frac{3}{2}$*. Let* ψ *fulfil Assumption 2.0.1 and let* $A(t)$ *be defined as in Section 2.1. Then for each* $T' < \infty$ *there exists a* $\mu > 0$ *such that*

$$\|(\mu + A(t))u\|_q$$

defines a family of equivalent norms for all $u \in \mathcal{D}(A)$ *and all* $t \leq T'$*. Here* $\| \cdot \|_q$ *denotes the norm of* $\mathrm{L}^q(\Omega_0)$*.*

Proof. As above, let $\tilde{A}(t) := \Phi(t)A_{\Omega(t)}\Phi(t)^{-1}$ denote the transformed Stokes operator. Then due to Section 1.5 we obtain that there exists a $c = c(t) > 0$ such that

$$\|\lambda u\|_q + \|\lambda^{\frac{1}{2}}\nabla u\|_q + \|\nabla^2 u\|_q \leq c(t)\|(\lambda + \tilde{A}(t))u\|_q$$

holds true for all $\lambda > 0$. We would like to prove, that the constant $c(t)$ can be chosen uniformly in $t \in [0, T']$ if we restrict ourselves to the case $\lambda \geq 1$. To prove that, let us compute first that

$$\|(\tilde{A}(t) - \tilde{A}(t'))u\|_q$$
$$\leq \|P(t) - P(t')\|_{\mathcal{L}(\mathrm{L}^q_\sigma(\Omega_0))}\Big\|\Big(\Delta + \sum_\alpha a_\alpha(t)\partial^\alpha\Big)u\Big\|_q$$
$$+ \|P(t')\|_{\mathcal{L}(\mathrm{L}^q_\sigma(\Omega_0))}\Big\|\sum_\alpha(a_\alpha(t) - a_\alpha(t'))\partial^\alpha u\Big\|_q$$
$$\leq c\|P(t) - P(t')\|_{\mathcal{L}(\mathrm{L}^q_\sigma(\Omega_0))}(1 + \sup_\alpha \|a_\alpha\|_{\mathrm{L}^\infty(\Omega_0)})\|u\|_{W^{2,q}(\Omega_0)}$$
$$+ c\|P(t')\|_{\mathcal{L}(\mathrm{L}^q_\sigma(\Omega_0))} \sup_\alpha \|a_\alpha(t) - a_\alpha(t')\|_{\mathrm{L}^\infty(\Omega_0)}\|u\|_{W^{2,q}(\Omega_0)}.$$

Let us define $\|u\|_{W^{2,q}_\lambda(\Omega_0)} := \lambda\|u\|_q + \lambda^{\frac{1}{2}}\|\nabla u\|_q + \|\nabla^2 u\|_q$. Thus we can estimate further

$$\|(\tilde{A}(t) - \tilde{A}(t'))u\|_q$$
$$\leq c\|P(t) - P(t')\|_{\mathcal{L}(\mathrm{L}^q_\sigma(\Omega_0))}(1 + \sup_\alpha \|a_\alpha\|_{\mathrm{L}^\infty(\Omega_0)})\|u\|_{W^{2,q}_\lambda(\Omega_0)}$$
$$+ c\|P(t')\|_{\mathcal{L}(\mathrm{L}^q_\sigma(\Omega_0))} \sup_\alpha \|a_\alpha(t) - a_\alpha(t')\|_{\mathrm{L}^\infty(\Omega_0)}\|u\|_{W^{2,q}_\lambda(\Omega_0)}$$

for all $\lambda \geq 1$ with a constant which does not depend on λ since

$$\| \cdot \|_{W^{2,q}} \leq \| \cdot \|_{W^{2,q}_\lambda}$$

if $\lambda \geq 1$. Due to the estimates in Lemma 2.1.7 and 2.1.8 we obtain that the map $t \mapsto A(t)$ is continuous as a map on

$$[0, \infty) \to \mathcal{L}(W_\lambda^{2,q}(\Omega_0) \cap W_0^{1,q}(\Omega_0) \cap L_\sigma^q(\Omega_0), L_\sigma^q(\Omega_0)).$$

Let $t, t_0 \in [0, T']$. Then we obtain

$$
\begin{aligned}
(\lambda + A(t))^{-1} &= \Big(\lambda + A(t_0) + (A(t) - A(t_0))\Big)^{-1} \\
&= (\lambda + A(t_0))^{-1}\Big(\mathrm{Id} + (A(t) - A(t_0))(\lambda + A(t))^{-1}\Big)^{-1}.
\end{aligned}
$$

Since $(\lambda + A(t_0))^{-1}\colon L_\sigma^q(\Omega_0) \to W_\lambda^{2,q}(\Omega_0)$ is uniformly bounded in $\lambda \geq 1$ and $t \mapsto A(t)$ is continuous as discussed above, we obtain that there exists a $\delta = \delta(t_0)$ such that

$$\Big\|(A(t) - A(t_0))(\lambda + A(t_0))^{-1}\Big\|_{\mathcal{L}(L_\sigma^q(\Omega_0))} < \frac{1}{2}$$

for all $t \in [t_0 - \delta(t_0), t_0 + \delta(t_0)] \cap [0, T']$ and all $\lambda \geq 1$. Hence, a Neumann series argument implies that

$$\|(\lambda + A(t))^{-1}\|_{\mathcal{L}(L_\sigma^q(\Omega_0), W_\lambda^{2,q}(\Omega_0))} \leq 2\|(\lambda + A(t_0))^{-1}\|_{\mathcal{L}(L_\sigma^q(\Omega_0), W_\lambda^{2,q}(\Omega_0))}$$

for all $t \in [t_0 - \delta(t_0), t_0 + \delta(t_0)] \cap [0, T']$ and all $\lambda \geq 1$. Thus, $(\lambda + A(t))^{-1}$ is uniformly bounded in a neighbourhood of t_0 and, since t_0 was arbitrary and $[0, T']$ is compact, we obtain that

$$\|(\lambda + A(t))^{-1}\|_{\mathcal{L}(L_\sigma^q(\Omega_0), W_\lambda^{2,q}(\Omega_0))}$$

is uniformly bounded for all $t \in [0, T']$. Thus, we have

$$(2.17) \qquad \lambda\|u\|_q + \lambda^{\frac{1}{2}}\|\nabla u\|_q + \|\nabla^2 u\|_q \leq c\|(\lambda + A(t))u\|_q$$

for all $u \in \mathcal{D}(A)$, all $\lambda \geq 1$ and all $t \in [0, T']$.

Let us now consider the operator

$$
\begin{aligned}
B(t)\colon W^{1,q}(\Omega_0) \cap L_\sigma^q(\Omega_0) &\to L_\sigma^q(\Omega_0), \\
u &\mapsto P(t)\sum_\beta b_\beta(t, \cdot)\partial^\beta u.
\end{aligned}
$$

Then the operator $B(t)$ is continuous for each t and due to Lemma 2.1.8 and the uniform boundedness of the coefficients b_β in the time- and spatial variable we obtain that

$$\sup_{t \leq T'} \|B(t)\|_{\mathcal{L}(W_0^{1,q}(\Omega_0) \cap L_\sigma^q(\Omega_0), L_\sigma^q(\Omega_0))} =: \|B\| < \infty.$$

Let $u \in \mathcal{D}(A)$ be arbitrary and let $t \in [0, T']$. Then we obtain for $\lambda \geq 1$

$$\|B(t')u\|_q \leq \|B\| \|u\|_{W^{1,q}(\Omega_0)}$$
$$\leq \frac{\|B\|}{\sqrt{\lambda}} \left(|\lambda| \|u\|_q + \sqrt{\lambda} \|\nabla u\|_q \right)$$
$$\leq c \frac{\|B\|}{\sqrt{\lambda}} \|(\lambda + \tilde{A}(t))u\|_q$$

for all $t \in [0, T']$ with the constant c from (2.17) which does not depend on t. Hence by choosing λ large enough we obtain

$$\|B(t')u\|_q \leq \frac{1}{2} \|(\lambda + \tilde{A}(t))u\|_q.$$

and the choice of λ is independent of t. Thus, we proved that

$$\left(\|(\lambda + \tilde{A}(t))u\|_q \right)$$

defines a family of equivalent norms and

(2.18) $$\frac{1}{2} \|(\lambda + \tilde{A}(t))u\|_q \leq \|(\lambda + A(t))u\|_q \leq 2 \|(\lambda + \tilde{A}(t))u\|_q.$$

This proves the assertion. $\qquad\square$

Remark 2.3.4. (i) Note that the main part in the proof of the last lemma is to show the uniformity of the resolvent estimate for $\lambda \geq 1$ of the transformed Stokes operator $\tilde{A}(t)$, $t \in [0, T']$. The idea to prove this is similar to a proof in [45].

(ii) Using the same ideas as above, the uniformity of the resolvent estimate can proved for all $\lambda \geq \delta$ for an arbitrary $\delta > 0$ and the constant in the resolvent estimate just depends on δ. Assuming that

$$A(t) \to A_\infty \in \mathcal{L}(W^{2,q}(\Omega_0) \cap W_0^{1,q}(\Omega_0) \cap L_\sigma^q(\Omega_0), L_\sigma^q(\Omega_0))$$

for an appropriate operator A_∞, one can prove the uniformity also for all $t \in [0, \infty)$ using the one-point compactification of $[0, \infty)$.

2.4 Maximal Regularity of the Transformed Stokes System

In this section we will prove the maximal L^p-regularity result to the Stokes system in a non-cylindrical time-space domain. The proof of the corresponding Theorem 2.1.10 is divided into two parts. First we will prove the result by assuming that $\|\phi - \mathrm{Id}\|_{C_b^{3,1}}$ is sufficiently small. We will see that in this case the result of M. Giga, Y. Giga, and H. Sohr will imply the validity of our main theorem. Since the motion of our domain comes to rest as time goes to infinity, this result for small data will imply that we get a maximal L^p-regularity result in a time interval $[T_0, \infty)$. In $[0, T_0)$ we can use the results of J. Saal to obtain the provided regularity result. Finally, Lemma 2.3.3 will be crucial to combine these results.

Let us start with an elementary lemma. In the following let $\mathcal{D}(\hat{A}_{\Omega_0})$ denote the domain of the Stokes operator on Ω_0 endowed with the homogeneous norm $\|A_{\Omega_0} u\|_q$.

Lemma 2.4.1. *Let $1 < q < \frac{3}{2}$ and let A_{Ω_0} denote the Stokes operator on the domain Ω_0.*

(i) *Let $a \colon \Omega_0 \to \mathbb{R}^{3,3}$ be a measurable function. Then the function*

$$\mathcal{D}(\hat{A}_{\Omega_0}) \to \mathrm{L}^q(\Omega_0), \ u \mapsto a\partial^\alpha u$$

is continuous provided that $a \in \mathrm{L}^r(\Omega_0, \mathbb{R}^{3,3})$ and

$$(r, |\alpha|) \in \left\{ \left(\frac{3}{2}, 0\right), (3, 1), (\infty, 2) \right\}.$$

(ii) *Assume that the map $a \in C^0([0,T]; \mathrm{L}^r(\Omega_0, \mathbb{R}^{3,3}))$. Then for any $u \in D(\hat{A}_{\Omega_0})$*

$$t \mapsto a(t)\partial^\alpha u \in C^0([0,T]; \mathrm{L}^q(\Omega_0, \mathbb{R}^3))$$

if the pair $(r, |\alpha|)$ fulfils the same relation as in part (i).

Proof. The proof is based on the inequalities (1.19), (1.20), and (1.23).

(i) Let $|\alpha| = 0$. Then

$$\|au\|_q \leq \|a\|_{\frac{3}{2}} \|u\|_{\frac{3q}{3-2q}} \leq c\|a\|_{\frac{3}{2}} \|A_{\Omega_0} u\|_q$$

due to the inequality (1.23). Note that the constant c depends on Ω_0 and on q. If $|\alpha| = 1$ we obtain

$$\|a\partial^\alpha u\|_q \leq \|a\|_3 \|\nabla u\|_{\frac{3q}{3-q}} \leq c\|a\|_3 \|A_{\Omega_0}^{\frac{1}{2}} u\|_{\frac{3q}{3-q}} \leq c\|a\|_3 \|A_{\Omega_0} u\|_q$$

using (1.23) and (1.20). Finally, if $|\alpha| = 2$ we may use (1.19) to obtain

$$\|a\partial^\alpha u\|_q \leq \|a\|_\infty \|\nabla^2 u\| \leq c\|a\|_\infty \|A_{\Omega_0} u\|_q.$$

(ii) The proof of (ii) is now clear using the estimates above. $\qquad\square$

As in the previous sections, let $\tilde{A}(t)$ denote the operator $\Phi(t) A_{\Omega(t)} \Phi(t)^{-1}$, i.e., $\tilde{A}(t)$ is the transformed Stokes operator. Then we have the following result.

Lemma 2.4.2. *Let $1 < q < \frac{3}{2}$ and let the motion of $\Omega(t)$ be described by a map ψ which fulfils Assumption 2.0.1.*

(i) *For each $t \in [0, \infty)$ the map*

$$t \mapsto A(t)\tilde{A}(0)^{-1}$$

extends to a bounded map on $L_\sigma^q(\Omega_0)$. The extension will also be denoted by $A(t)\tilde{A}(0)^{-1}$

(ii) *We have $t \mapsto A(t)\tilde{A}(0)^{-1} \in C^0([0, \infty); \mathcal{L}(L_\sigma^q(\Omega_0)))$.*

(iii) *The map $\tilde{A}(t)\tilde{A}(0)^{-1}$ has the same properties as $A(t)\tilde{A}(0)^{-1}$ listed in (i) and (ii).*

Proof. The proof is based on the last Lemma 2.4.1.

(i) Since

$$A(t) - \tilde{A}(0) = (P(t) - P(0))(-\Delta) + P(t)\Big(-\sum_{\alpha} a_{\alpha}(t)\partial^{\alpha} + \sum_{\beta} b_{\beta}(t)\partial^{\beta}\Big)$$

we obtain

$$
\begin{aligned}
\|A(t)u\|_q &\leq \|(A(t) - \tilde{A}(0))u\|_q + \|\tilde{A}(0)u\|_q \\
&\leq \|P(t) - P(0)\|_{\mathcal{L}(L^q_\sigma(\Omega_0))}\|\Delta u\|_q \\
&\quad + c\sup_{\alpha,\beta}(\|a_\alpha(t)\|_\infty + \|b_\beta(t)\|_\infty)\|\tilde{A}(0)u\|_q \\
&\leq c\big(...\big)\|\tilde{A}(0)u\|_q.
\end{aligned}
$$

and the term in the brackets is given by

$$\big(...\big) = \Big(\|P(t) - P(0)\|_{\mathcal{L}(L^q_\sigma(\Omega_0))} + \sup_{\alpha,\beta}(\|a_\alpha(t)\|_\infty + \|b_\beta(t)\|_\infty)\Big).$$

Note that we used here that a_α, b_β have compact support in $B_R(0)$, see Lemma 2.1.7, and hence we obtain $\|a_\alpha\|_r \leq c\|a_\alpha\|_\infty$ with a constant just depending on $r \in [1,\infty)$ and the support of a_α. The final estimate is valid due to Lemma 2.4.1. Thus, we proved that for all $u \in \mathcal{R}(\tilde{A}(0))$ the estimate

$$\|A(t)\tilde{A}(0)^{-1}u\|_q \leq c\|u\|_q$$

holds, and since $\tilde{A}(0)$ has dense range, the operator $A(t)\tilde{A}(0)^{-1}$ extends to a continuous operator on $L^q_\sigma(\Omega)$.

(ii) Note that

$$
\begin{aligned}
A(t) - A(s) &= (P(t) - P(s))\Big(-\Delta - \sum_{\alpha} a_{\alpha}(t)\partial^{\alpha} + \sum_{\beta} b_{\beta}(t)\partial^{\beta}\Big) \\
&\quad + P(s)\Big(\sum_{\alpha} \tilde{a}_{\alpha}(t,s)\partial^{\alpha}\Big),
\end{aligned}
$$

where

$$\tilde{a}_\alpha(t,s) := -(a_\alpha(t) - a_\alpha(s)) + (b_\alpha(t) - b_\alpha(s))$$

using the convention that $b_\alpha = 0$ if $|\alpha| = 2$. Since $\|P(t) - P(s)\|_{\mathcal{L}(L^q_\sigma(\Omega_0))}$ is continuous in s and t, see Lemma 2.1.8, and $(t,s) \mapsto \tilde{a}_\alpha(t,s)$ is continuous for all α, see Lemma 2.1.7, we obtain the desired result with the same estimates as in (i).

(iii) This proof can be done as in part (i) and (ii) by neglecting the terms corresponding to b_β. □

In fact, we can extract a little bit more from the proof of the last result. The estimates above imply that since the coefficients $a_\alpha(t)$ and $b_\alpha(t)$ are converging to some coefficients as $t \to \infty$ in $L^\infty(\Omega_0)$ and $P(t)$ is also converging to some projection as $t \to \infty$ in $\mathcal{L}(L^q(\Omega))$, see the estimate in Lemma 2.1.8, we obtain that the operator $t \mapsto A(t)\tilde{A}(0)^{-1}$ is converging as $t \to \infty$. Due to a compactness argument, it follows that

$$\|A(t)u\|_q \le c\|\tilde{A}(0)u\|_q$$

for all $u \in \mathcal{D}(A)$ with a constant c which can be chosen uniformly in t. With the following Lemma, we would like to prove that also a reverse inequality with $A(t)$ replaced by $\tilde{A}(t)$ holds true with some uniform constant.

Lemma 2.4.3. *Let* $1 < q < \frac{3}{2}$ *and let the motion of* $\Omega(t)$ *be described by a map* ψ *which fulfils Assumption 2.0.1.*

(i) *For each* $t \in (0, \infty)$ *there exists a constant* $c = c(t)$ *such that*

$$(2.19) \qquad \|\nabla^2 u\|_q \le c(t)\|\tilde{A}(t)u\|_q$$

for all $u \in \mathcal{D}(A)$. *This result also holds true for the transformed Stokes operator* A_∞.

(ii) *The constant* $c(t)$ *in part (i) can be chosen uniformly in* t. *Especially, there exists* $c_1, c_2 > 0$ *such that*

$$(2.20) \qquad c_1\|\tilde{A}(t)u\|_q \le \|\tilde{A}(\tau)u\|_q \le c_2\|\tilde{A}(t)u\|_q$$

holds true for all $t, \tau \in [0, \infty)$.

Proof. (i) Let $t \in (0, \infty)$ be arbitrary and let $u \in \mathcal{D}(A_{\Omega_0})$. Then we obtain

$$\|\nabla^2 u\|_{L^q(\Omega_0)} \le c\|\Delta u\|_{L^q(\Omega_0)}$$
$$\le c\|\Phi(t)^{-1}\Delta\Phi(t)v\|_{L^q(\Omega(t))}$$

with $v := \Phi(t)^{-1}u$, where we used the isomorphism property of $\Phi(t)$ in L^q, see Lemma 2.1.1 (i). Thus, we obtain

$$\|\nabla^2 u\|_q \le c \left\|\left(\Delta + \sum_\alpha a_\alpha^t(0, \cdot)\partial^\alpha\right)v\right\|_{L^q(\Omega(t))}$$

with coefficient a_α^t as in Lemma 2.1.11 and this yields

$$\|\nabla^2 u\|_q \le c\|A_{\Omega(t)}v\|_{L^q(\Omega(t))}$$

using the same kind of estimates as in Lemma 2.4.1. Using Lemma 2.1.1 we obtain

$$\|\nabla^2 u\|_q \le c\|\Phi(t)A_{\Omega(t)}\Phi(t)^{-1}u\|_{L^q(\Omega_0)}$$
$$= c\|\tilde{A}(t)u\|_{L^q(\Omega_0)}.$$

Note that at least the constant in the second last line cannot be chosen *a priori* to be independently of t since the domain depending Sobolev estimates (1.23) are used. The proof for the case $t = \infty$ can be done exactly in the same way.

(ii) Let $t_0 \in (0, \infty)$ be arbitrary. Due to the estimate in (i) we obtain for $u \in \mathcal{D}(A_{\Omega_0})$ that

$$\|\tilde{A}(0)u\|_q \le c\|\nabla^2 u\|_q \le c\|\tilde{A}(t_0)u\|_q,$$

and thus Lemma 2.4.2 implies that

$$t \mapsto \tilde{A}(t)\tilde{A}(t_0)^{-1} = \tilde{A}(t)\tilde{A}(0)^{-1}\tilde{A}(0)\tilde{A}(t_0)^{-1} \in C^0([0, \infty); \mathcal{L}(L^q(\Omega_0))).$$

This yields the existence of a neighbourhood $\mathcal{U}(t_0)$ of t_0 such that

$$\|(\tilde{A}(t) - \tilde{A}(t_0))u\|_q \le \frac{1}{4}\|\tilde{A}(t_0)u\|_q$$

for all $t \in \mathcal{U}(t_0)$ and therefore the estimates

$$\|\tilde{A}(t)u\| \le \|\tilde{A}(t_0)u\|_q + \|(\tilde{A}(t) - \tilde{A}(t_0))u\|_q \le \frac{5}{4}\|\tilde{A}(t_0)u\|_q$$

and

$$\|\tilde{A}(t)u\|_q \geq \|\tilde{A}(t_0)u\|_q - \|(\tilde{A}(t) - \tilde{A}(t_0))u\|_q \geq \frac{3}{4}\|\tilde{A}(t_0)u\|_q$$

imply that for all $t \in \mathcal{U}(t)$ the constant in the estimate (2.19) can be chosen uniformly. Note that the same holds true for $t_0 = \infty$.

Finally, since $[0,\infty]$ is compact, *i.e.*, considering the one-point compactification of $[0,\infty)$, we can cover this interval by finitely many $\mathcal{U}(t_i)$, $1 \leq i \leq n$, $t_i \in [0,\infty]$, to obtain that the constant c in estimate (2.19) can be chosen uniformly in t, *e.g.*,

$$c := \frac{5}{4}\max\{c(t_i) \mid 1 \leq i \leq n\}$$

is an adequate choice for c. Finally, due to

$$\|A_{\Omega_0}u\|_q \leq \bar{c}\|\nabla^2 u\|_q \leq \tilde{c}\|\tilde{A}(t)u\|_q$$

for all $u \in \mathcal{D}(A_{\Omega_0})$ with a constant $\bar{c}$ which depends on Ω_0, but not on t, and a constant $\tilde{c}$ chosen uniformly in t, we obtain the existence of constants c_1, c_2 as in (2.20). $\qquad\square$

Let us now start to prove a maximal regularity result to the non-autonomous system (2.9). The main idea in the proof is to apply Theorem 2.2.2, and to prove its applicability we will use a perturbation argument. The main difficulty here is that even if $b_\beta(t,\cdot) \to 0$ as $t \to \infty$, b_β might be large for small t. Let us first avoid this difficulty by assuming that $\|\phi - \mathrm{Id}\|_{C_b^{3,1}}$ is sufficiently small. The following theorem will be applied on an interval $[T_0, \infty)$ for a sufficient large T_0.

Theorem 2.4.4. *Let $1 < q < \frac{3}{2}$ and $1 < p < \infty$. Let ψ fulfil Assumption 2.0.1 and let $A(t)$ be defined as above. Then there exists a parameter $\delta > 0$, which depends only on p, q, with the following property: Assume that $\|\phi - \mathrm{Id}\|_{C_b^{3,1}} < \delta$, then the system (2.9) has maximal L^p-regularity, i.e., for all $u_0 \in \mathscr{J}_{A_{\Omega_0}}^p$ and all $f \in \mathrm{L}^p(0,\infty;\mathrm{L}_\sigma^q(\Omega_0))$ there exists a unique solution to (2.9). Furthermore, there exists $c > 0$ which does not depend on f, u_0 such that the estimate*

$$\|u_t\|_{q,p} + \|A(\cdot)u\|_{q,p} \leq c\left(\|u_0\|_{\mathscr{J}_{A(0)}^p} + \|f\|_{q,p}\right)$$

holds true.

Proof. Let us prove that the assumptions of Theorem 2.2.2 are fulfilled. Therefore, note that the operators $A(t)$ are defined on the dense subspace $W^{2,q}(\Omega_0) \cap W_0^{1,q}(\Omega_0) \cap L_\sigma^q(\Omega) \subset L_\sigma^q(\Omega)$.

Furthermore, due to Lemma 2.1.7 we obtain that

$$\|a_\alpha(\cdot, t)\|_\infty + \|b_\beta(\cdot, t)\|_\infty \le c\|\phi - \mathrm{Id}\|_{C_b^{3,1}} \le c\delta$$

for all $t \in [0, \infty)$. Hence if δ is small enough, we obtain with Lemmas 2.4.1 - 2.4.3 that

$$\|(A(t) - A(\tau))u\|_q \le c\delta\|\tilde{A}_\infty u\|_q$$

for all $t, \tau \in [0, \infty]$ and $u \in \mathcal{D}(A_{\Omega_0})$.

Hence, a perturbation result for sectorial operators, see [10, 1.5 Theorem], implies that the operators $A(t)$ are sectorial since they are perturbations of A_∞; therefore, the operators $A(t)$ have dense range and the resolvent estimate in Theorem 2.2.2 (ii), is fulfilled.

In addition, we obtain

$$(2.21) \qquad \|A(\tau)u - \tilde{A}(\tau)u\|_q \le c\delta\|\tilde{A}_\infty u\|_q \le c\delta\|\tilde{A}(\tau)u\|_q$$

due to Lemma 2.1.7 (ii) and (2.20), and hence

$$\begin{aligned}
\|A(t)u\|_q &\le \|(A(t) - A(\tau))u\|_q + \|A(\tau)u\|_q \\
&\le c\delta\|\tilde{A}_\infty u\|_q + (1 + c\delta)\|\tilde{A}(\tau)\|_q \\
&\le c\|A(\tau)u\|_q,
\end{aligned}$$

and thus the assumptions (iii) in Theorem 2.2.2 is fulfilled.

Furthermore, since the property of maximal regularity is invariant under similarity transforms, we obtain that for all t the operator $\tilde{A}(t)$ has maximal regularity. Due to the characterisation of maximal regularity in Lemma 1.3.13 and the perturbation result of Lemma 1.3.11, we obtain that the operator $A(t)$ has maximal regularity. Finally, since

$$\|(A(t) - \tilde{A}_\infty)u\|_q \le c\delta\|\tilde{A}_\infty u\|_q$$

due to the estimates of $P(\cdot)$ and the coefficients a_α and b_β, we obtain by Proposition 1.3.14 that the constants in the maximal regularity estimate for $A(t)$ can be chosen uniformly. Thus, also assumption (iv) in Theorem 2.2.2 is fulfilled. Finally, due to the continuity of $t \mapsto A(t)\tilde{A}(0)^{-1}$ and the norm equivalence uniformly in t of $\|A(t)u\|_q$ and $\|\tilde{A}(0)u\|_q$ we obtain that also assumption (v) in Theorem 2.2.2 is satisfied. This yields the desired result. $\qquad\square$

Now we are able to prove our main Theorem 2.1.10. Let us remark that in the following proof one of the crucial steps is to consider the shifted operators $\lambda + A(t)$ for t in a finite time interval. As discussed in Remark 1.3.16, this will be the reason why we have to consider the norm of $\mathscr{I}^p_{1+\tilde{A}(0)}$ on the right-hand side of (2.10).

Proof of Theorem 2.1.10. Note that we can prove the existence of a solution to the non-autonomous system in the same way as in the proof of Theorem 2.2.2. Hence we just have to prove the L^p-L^q-estimate. Therefore, let us note that due to the last Theorem 2.4.4 we obtain the existence of a $T_0 \geq 0$ such that the L^p-L^q-estimate holds on the time interval $[T_0, \infty)$, hence we obtain the estimate

$$(2.22) \quad \begin{aligned} &\|u_t\|_{L^p(T_0,\infty;L^q_\sigma(\Omega_0))} + \|A(\cdot)u\|_{L^p(T_0,\infty;L^q_\sigma(\Omega_0))} \\ &\leq c\Big(\|f\|_{L^p(T_0,\infty;L^q_\sigma(\Omega_0))} + \|u(T_0)\|_{\mathscr{I}^p_{A(T_0)}}\Big). \end{aligned}$$

Let us consider the non-autonomous system on the finite time interval $[0, T_0)$. Note that for every $\lambda > 0$ we can apply the result of J. Saal, see Theorem 2.3.1, to the operator $A_\lambda(t) := \lambda + A(t)$ to obtain a L^p-L^q-estimate on the time interval $[0, T_0)$ to the operator $A_\lambda(t)$. Let us choose $\lambda > 0$ such that

$$\Big(\|A_\lambda(t)u\|_q\Big)_{0 \leq t \leq T_0}$$

defines a family of equivalent norms for all $u \in \mathcal{D}(A)$, which is possible due to Lemma 2.3.3. Using Theorem 2.3.1 we obtain the estimate

$$(2.23) \quad \begin{aligned} &\|u_t\|_{L^p(0,T_0;L^q_\sigma(\Omega_0))} + \|A(\cdot)u\|_{L^p(0,T_0;L^q_\sigma(\Omega_0))} \\ &\leq c\Big(\|f\|_{L^p(0,T_0;L^q_\sigma(\Omega_0))} + \|u_0\|_{\mathscr{I}^p_{1+\tilde{A}(0)}}\Big) \end{aligned}$$

with a constant c which depends in particular on λ and T_0. Combining (2.22) and (2.23) we obtain the estimate

$$(2.24) \quad \begin{aligned} &\|u_t\|_{L^p(0,\infty;L^q_\sigma(\Omega_0))} + \|A(\cdot)u\|_{L^p(0,\infty;L^q_\sigma(\Omega_0))} \\ &\leq c\Big(\|f\|_{L^p(0,\infty;L^q_\sigma(\Omega_0))} + \|u_0\|_{\mathscr{I}^p_{1+\tilde{A}(0)}} + \|u(T_0)\|_{\mathscr{I}^p_{A(T_0)}}\Big), \end{aligned}$$

and hence we just have to estimate the term $\|u(T_0)\|_{\mathscr{I}^p_{A(T_0)}}$ on the right hand side of (2.24). In Lemma 1.3.3 we proved that the space $\mathscr{I}^p_{A(T_0)}$ coincides with equivalent norms with the space $\mathscr{F}^p_{A(T_0)}$. Let us consider the function

$$w \colon t \mapsto \begin{cases} u(T_0 - t), & t \leq T_0, \\ e^{-(t-T_0)A(0)}u_0, & t \geq T_0, \end{cases}$$

which satisfies $w(0) = u(T_0)$. Hence we can estimate as follows

$$\|u(T_0)\|_{\mathscr{I}^p_{A(T_0)}} \leq c\|u(T_0)\|_{\mathscr{F}^p_{A(T_0)}} \leq c\|w\|_{W^p_{A(T_0)}}$$
$$\leq c\Big(\|w_t\|_{\mathrm{L}^p(0,T_0;\mathrm{L}^q_\sigma(\Omega_0))} + \|A(T_0)w\|_{\mathrm{L}^p(0,T_0;\mathrm{L}^q_\sigma(\Omega_0))}$$
$$+ \|w_t\|_{\mathrm{L}^p(T_0,\infty;\mathrm{L}^q_\sigma(\Omega_0))} + \|A(T_0)w\|_{\mathrm{L}^p(T_0,\infty;\mathrm{L}^q_\sigma(\Omega_0))}\Big)$$
$$\leq c\Big(\|w_t\|_{\mathrm{L}^p(0,T_0;\mathrm{L}^q_\sigma(\Omega_0))} + \|A_\lambda(T_0)w\|_{\mathrm{L}^p(0,T_0;\mathrm{L}^q_\sigma(\Omega_0))}$$
$$+ \|w_t\|_{\mathrm{L}^p(T_0,\infty;\mathrm{L}^q_\sigma(\Omega_0))} + \|\tilde{A}(0)w\|_{\mathrm{L}^p(T_0,\infty;\mathrm{L}^q_\sigma(\Omega_0))}\Big)$$

due to the resolvent estimate and Lemma 2.4.2. Thus we get

$$\|u(T_0)\|_{\mathscr{I}^p_{A(T_0)}} \leq c\Big(\|u_t\|_{\mathrm{L}^p(0,T_0;\mathrm{L}^q_\sigma(\Omega_0))} + \|A_\lambda(t)u(t)\|_{\mathrm{L}^p(0,T_0;\mathrm{L}^q_\sigma(\Omega_0))} + \|u_0\|_{\mathscr{I}^p_{1+\tilde{A}(0)}}\Big)$$

by the equivalence of the norms $\left(\|A_\lambda(t)u\|_q\right)_t$, and finally using (1.13) we obtain

$$\|u(T_0)\|_{\mathscr{I}^p_{A(T_0)}} \leq c\Big(\|u_t\|_{\mathrm{L}^p(0,T_0;\mathrm{L}^q_\sigma(\Omega_0))} + \|A(t)u(t)\|_{\mathrm{L}^p(0,T_0;\mathrm{L}^q_\sigma(\Omega_0))} + \|u_0\|_{\mathscr{I}^p_{1+\tilde{A}(0)}}\Big)$$
$$\leq c\Big(\|u_0\|_{\mathscr{I}^p_{1+\tilde{A}(0)}} + \|f\|_{\mathrm{L}^p(0,T_0;\mathrm{L}^q_\sigma(\Omega_0))}\Big)$$

with (2.22). Hence we have proved the maximal L^p-regularity result to the non- autonomous system. □

Finally, we have to prove that Theorem 2.1.10 implies Theorem 2.0.2.

Proof of Theorem 2.0.2. As discussed after Theorem 2.1.10, the existence of a strong solution is already proven by Theorem 2.1.10 and we just have to prove the estimate (2.11) to finalize the proof. Hence let us start to estimate

$$\|(\partial_t + B(\cdot))u\|_{\mathrm{L}^p(0,\infty;\mathrm{L}^q_\sigma(\Omega_0))} + \|\tilde{A}(\cdot)u\|_{\mathrm{L}^p(0,\infty;\mathrm{L}^q_\sigma(\Omega_0))}$$
$$\leq \|\partial_t u\|_{\mathrm{L}^p(0,\infty;\mathrm{L}^q_\sigma(\Omega_0))} + \|B(\cdot)u\|_{\mathrm{L}^p(0,\infty;\mathrm{L}^q_\sigma(\Omega_0))} + \|\tilde{A}(\cdot)u\|_{\mathrm{L}^p(0,\infty;\mathrm{L}^q_\sigma(\Omega_0))}$$
$$\leq \|\partial_t u\|_{\mathrm{L}^p(0,T_0;\mathrm{L}^q_\sigma(\Omega_0))} + \|B(\cdot)u\|_{\mathrm{L}^p(0,T_0;\mathrm{L}^q_\sigma(\Omega_0))} + \|\tilde{A}(\cdot)u\|_{\mathrm{L}^p(0,T_0;\mathrm{L}^q_\sigma(\Omega_0))}$$
$$+ \|\partial_t u\|_{\mathrm{L}^p(T_0,\infty;\mathrm{L}^q_\sigma(\Omega_0))} + \|B(\cdot)u\|_{\mathrm{L}^p(T_0,\infty;\mathrm{L}^q_\sigma(\Omega_0))} + \|\tilde{A}(\cdot)u\|_{\mathrm{L}^p(T_0,\infty;\mathrm{L}^q_\sigma(\Omega_0))}$$

with the same T_0 as in the proof of Theorem 2.1.10.

Choosing $\lambda > 0$ large enough, we obtain using (2.18) that

$$\|B(\cdot)u\|_{\mathrm{L}^p(0,T_0;\mathrm{L}^q_\sigma(\Omega_0))} \leq c\|(\lambda + \tilde{A}(\cdot))u\|_{\mathrm{L}^p(0,T_0;\mathrm{L}^q_\sigma(\Omega_0))}$$
$$\leq c\|(\lambda + A(\cdot))u\|_{\mathrm{L}^p(0,T_0;\mathrm{L}^q_\sigma(\Omega_0))}.$$

In addition, due to (2.21) we also obtain that

$$\|B(\cdot)u\|_{\mathrm{L}^p(T_0,\infty;\mathrm{L}^q_\sigma(\Omega_0))} \leq c\|A(\cdot)u\|_{\mathrm{L}^p(T_0,\infty;\mathrm{L}^q_\sigma(\Omega_0))}.$$

This proves that we can estimate with the right-hand side of a maximal L^p-regularity estimate. $\qquad\square$

CHAPTER 3

Navier–Stokes Equations in a Non-cylindrical Time-Space Domain

---

In this chapter we would like to prove an existence result for strong solutions to the Navier–Stokes equations in non-cylindrical time space domains introduced in the Introduction and in Chapter 2. Hence we will consider the system

$$(3.1) \qquad \begin{aligned} v_t - \Delta v + v \cdot \nabla v + \nabla p &= f \text{ in } Q, \\ \operatorname{div} v &= 0 \text{ in } Q, \end{aligned}$$

completed with (non-homogeneous) Dirichlet boundary data and initial condition $v(0) = v_0$ on Ω_0. Recall that Q denotes the non-cylindrical time space domain $Q := \bigcup_t \Omega(t) \times \{t\}$ and Γ the boundary w.r.t. the spatial variable $\bigcup_{t>0} \partial\Omega(t) \times \{t\}$ introduced in Chapter 2.

Using the same transformation as before, we can transform the system (3.1) to the non-autonomous system

$$(3.2) \qquad \begin{aligned} u_t + A(t)u + P(t)(u \cdot \nabla^\phi(t)u) &= f \text{ in } (0, \infty) \times \Omega_0, \\ u(0) &= u_0 \text{ in } \Omega_0 \end{aligned}$$

applying the transformed Helmholtz projection; the transformed gradient will be defined in (3.3) in the Proof of Lemma 3.2.1. Concerning the boundary data, we assume that the transformed system (3.2) is endowed with homogeneous Dirichlet boundary data since the domain of $A(t)$ is a subspace of $W_0^{1,q}(\Omega_0)$. As we will discuss in the next chapter, one can generalize the system easily to the case of non-homogeneous boundary data by constructing a solution $u = \bar{u} + \tilde{u}$, where $\tilde{u}$ denotes an extension of the boundary data and $\bar{u}$ is a solution to a Navier–Stokes type equation with homogeneous Dirichlet boundary data. We will discuss this way to solve the Navier–Stokes equations carefully for weak solutions in the next Chapter and emphasize the idea to deal with non-homogeneous boundary data later on.

3.1 Preliminary Results to the Non-linear Problem

As in many situations, the idea to construct a solution to the non-linear problem is to rewrite the system into a fixed point problem and prove that we can apply Banach's Fixed Point Theorem to obtain the existence of a solution.

Lemma 3.1.1. *Let X be a Banach space and let $\mathcal{F}\colon X \to X$ be a continuous operator such that*

$$\|\mathcal{F}(x)\|_X \le \alpha(\|x\|_X + \beta)^2$$

for all $x \in X$ and some $\alpha, \beta > 0$ such that $4\alpha\beta < 1$. Furthermore, assume that

$$\|\mathcal{F}(x) - \mathcal{F}(y)\|_X \le \alpha(\|x\|_X + \|y\|_X + 2\beta)\|x - y\|_X$$

for all $x, y \in X$. Then there exists an $r > 0$ such that $\mathcal{F}(\overline{B_r}) \subset B_r$ and there exists a unique fixed point of $\mathcal{F}$ in B_r.

Proof. Let

$$y_1 := \frac{1}{2\alpha}(1 - \sqrt{1 - 4\alpha\beta}) = \frac{1}{2\alpha}\frac{4\alpha\beta}{1 + \sqrt{1 - 4\alpha\beta}} = \frac{2\beta}{1 + \sqrt{1 - 4\alpha\beta}}$$

denote the smallest root of $y - \alpha y^2 - \beta$. Then we obtain $y_1 \in (\beta, 2\beta)$ and let us define $r := y_1 - \beta$. Thus we get

$$\|\mathcal{F}(x)\|_X \le \alpha(\|x\|_X + \beta)^2 \le \alpha y_1^2 = y_1 - \beta = r$$

for all $x \in \overline{B_r}$. Furthermore, we have

$$\|\mathcal{F}(x) - \mathcal{F}(y)\|_X \leq 2\alpha y_1 \|x - y\|_X \leq 4\alpha\beta\|x - y\|_X$$

and hence $\mathcal{F}$ is a strict contraction on $\overline{B_r}$. Finally, Banach's Fixed Point Theorem implies that $\mathcal{F}$ has a unique fixed point in $\overline{B_r}$. $\qquad\square$

Note that the last statement does not prove that the fixed point of the mapping $\mathcal{F}$ is unique since there might be a second fixed point in $\overline{B_r}^c$.

3.2 Application to the Navier–Stokes Equations

The crucial point to prove an existence result to a non-linear differential equation using a maximal regularity result is to estimate the non-linear term in a suitable way. This will be done in the Lemma 3.2.1.

Let us recall the notation $W_{\tilde{A}(0)}^p$, introduced in Section 1.3, of all measurable function $u\colon [0,\infty) \to \mathrm{L}_\sigma^q(\Omega_0)$ such that $u_t, \tilde{A}(0)u \in \mathrm{L}^p(0,\infty;\mathrm{L}_\sigma^q(\Omega_0))$ endowed with the canonical norm.

Lemma 3.2.1. *Let* $1 < q < \frac{3}{2}$ *and* $1 < p < \infty$ *such that* $\frac{2}{p} + \frac{3}{q} = 3$. *Let* $w_1, w_2 \in W_{\tilde{A}(0)}^p$ *such that* $w_1(0) = w_2(0) = 0$. *Then there exists a constant* $c > 0$ *such that*

$$\|P(\cdot)\big(w_1 \cdot \nabla^\phi(\cdot)w_2\big)\|_{\mathrm{L}^p(0,\infty;\mathrm{L}_\sigma^q(\Omega_0))} \leq c\|w_1\|_{W_{\tilde{A}(0)}^p}\|w_2\|_{W_{\tilde{A}(0)}^p}.$$

Of course, if the time interval is finite the condition on p can be weakened to $\frac{2}{p} + \frac{3}{q} \leq 3$ due to the embeddings of Lebesgue-Bochner spaces on space with finite measure.

Proof. Due to Lemma 2.1.8 we get that the family of projections $P(t)$ on $\mathrm{L}^q(\Omega_0)$ is uniformly bounded. Furthermore, since

$$(3.3) \qquad \nabla^\phi(t) = (\nabla\phi(t))^{-1}(\nabla\phi(t))^{-T}\nabla =: \zeta(t)\nabla$$

for some uniformly bounded matrix $\zeta(t)$, we obtain the desired result if we prove an appropriate estimate to $\|w_1 \cdot \nabla w_2\|_{q,p}$. Due to Theorem 1.4.4

we can apply the Mixed Derivative Theorem 1.4.1 to the time derivative ∂_t and the natural extension of the Stokes operator $\tilde{A}(0)$ to the domain $L^p(0, \infty; \mathcal{D}(A))$. Let $\alpha = \frac{1}{2p}$. Then we obtain

$$\|w_1 \cdot \nabla w_2\|_{q,p} \le \|w_1\|_{q_1,2p} \|\nabla w_2\|_{q_2,2p}$$

with $\frac{1}{q_1} = \frac{1}{q} - \frac{2}{3}(1 - \alpha)$ and $\frac{1}{q_2} = \frac{1}{q} - \frac{1}{3}(1 - 2\alpha)$; the Hölder inequality can be applied due to

$$\frac{1}{q_1} + \frac{1}{q_2} = \frac{2}{q} - 1 + \frac{4}{3}\alpha = \frac{2}{q} - 1 + \frac{1}{3}\left(3 - \frac{3}{q}\right) = \frac{1}{q},$$

and we can continue with

$$\|w_1 \cdot \nabla w_2\|_{q,p} \le c\|A^{1-\alpha} w_1\|_{q,2p} \|A^{1-\alpha} w_2\|_{q,2p}$$

using the estimate (1.23), which is applicable since $\frac{1}{q_2} = \frac{1}{2q} + \frac{1}{6} \ge \frac{1}{2}$ and thus $q_2 < 3$. Note that we simplified the notation by $A = \tilde{A}(0)$. Moreover

$$\|w_1 \cdot \nabla w_2\|_{q,p} \le c\|\partial_t^\alpha A^{1-\alpha} w_1\|_{q,p} \|\partial_t^\alpha A^{1-\alpha} w_2\|_{q,p}.$$

The last step is valid due to Proposition 1.4.5. The Mixed Derivative Theorem 1.4.1 now yields the result. $\qquad\square$

From now on, let $0 < T \le \infty$ and let $\bar{u} \in L^p(0, T; L^q_\sigma(\Omega_0))$. We consider the operator

$$(3.4) \qquad \begin{aligned} \mathcal{F}_{\bar{u}} \colon\ & L^p(0, T; L^q_\sigma(\Omega_0)) \to L^p(0, T; L^q_\sigma(\Omega_0)), \\ & g \mapsto -P(\cdot)\big((L(\cdot)^{-1}g + \bar{u}) \cdot \nabla^\phi(\cdot)(L(\cdot)^{-1}g + \bar{u})\big). \end{aligned}$$

Here $L(\cdot)$ denotes the operator $\partial_t + A(\cdot)$ and $L(\cdot)^{-1}$ denotes the solution operator to

$$L(\cdot)u = f, \ u(0) = 0.$$

Furthermore, let $L := \partial_t + A(0)$. Let us start with the following, elementary calculation.

Lemma 3.2.2. *Let $1 < q < \frac{3}{2}$, $1 < p < \infty$, and let $\bar{u} \in W^p_{A(0)}$. Furthermore, let $g \in L^p(0, T; L^q_\sigma(\Omega_0))$. Then*

$$u := L(\cdot)^{-1}g + \bar{u}$$

is a solution to

(3.5) $\qquad u_t + A(\cdot)u + P(t)\big(u \cdot \nabla^\phi(\cdot)u\big) = \bar{u}_t + A(\cdot)\bar{u}, \ u(0) = \bar{u}(0)$

if and only if g is a fixed point of $\mathcal{F}_{\bar{u}}$.

Please note that this Lemma does not require any condition on the integrability in time $p \in (1, \infty)$ or on the time interval $(0, T)$, $0 < T \le \infty$.

Proof. Let g be a fixed point of $\mathcal{F}_{\bar{u}}$. Then

$$
\begin{aligned}
u_t + A(\cdot)u + P(t)\big(u \cdot \nabla^\phi(\cdot)u\big) &= L(\cdot)(L(\cdot)^{-1}g) + L(\cdot)\bar{u} \\
&\quad + P(t)\big((L(\cdot)^{-1}g + \bar{u}) \cdot \nabla^\phi(\cdot)(L(\cdot)^{-1}g + \bar{u})\big) \\
&= L(\cdot)\bar{u}
\end{aligned}
$$

and hence we proved the first implication. If otherwise u is a solution to (3.5), then the calculation above implies that

$$L(\cdot)(L(\cdot)^{-1}g) = -P(t)\big((L(\cdot)^{-1}g + \bar{u}) \cdot \nabla^\phi(\cdot)(L(\cdot)^{-1}g)\big),$$

thus $g = L(\cdot)(u - \bar{u})$ is a fixed point of $\mathcal{F}_{\bar{u}}$. $\qquad \square$

Due to the last lemma it is clear that the existence of a solution to (3.2) can be reduced to the linearized problem discussed in the previous chapter and the existence of a fixed point of the operator $\mathcal{F}$. Let us start with an existence result for short times.

Theorem 3.2.3. *Let $1 < q < \frac{3}{2}$ and let $1 < p < \infty$ such that $\frac{2}{p} + \frac{3}{q} \le 3$. Let $u_0 \in \mathscr{I}^p_{1+\tilde{A}(0)}$ and let $f \in L^p(0, T; L^q_\sigma(\Omega_0))$, $0 < T < \infty$. Then there exists $0 < T' \le T$ such that there exists a unique solution u to (3.2) in the time interval $(0, T')$.*

Proof. Let us endow the space of initial values with the norm $\mathscr{F}^p_{1+\tilde{A}(0)}$ defined in the preliminaries. Note that this norm depends on T' and tends to 0 as $T' \to 0$, see Remark 1.3.6. Let $\bar{u}$ be a solution to the linearized problem with data f and u_0. Due to the results in the previous chapter we obtain that

$$\|\bar{u}_t\|_{\mathrm{L}^p(0,T';\mathrm{L}^q_\sigma(\Omega_0))} + \|A(\cdot)\bar{u}\|_{\mathrm{L}^p(0,T';\mathrm{L}^q_\sigma(\Omega_0))} \leq c\Big(\|f\|_{\mathrm{L}^p(0,T';\mathrm{L}^q_\sigma(\Omega_0))} + \|u_0\|_{\mathscr{F}^p_{1+\tilde{A}(0)}}\Big),$$

where the constant c can be chosen independently of T'. We obtain for any large $\lambda \geq 1$ that

$$
\begin{aligned}
&\|\bar{u}_t\|_{\mathrm{L}^p(0,T';\mathrm{L}^q_\sigma(\Omega_0))} + \|\tilde{A}(0)\bar{u}\|_{\mathrm{L}^p(0,T';\mathrm{L}^q_\sigma(\Omega_0))} \\
&\leq c\Big(\|\bar{u}_t\|_{\mathrm{L}^p(0,T';\mathrm{L}^q_\sigma(\Omega_0))} + \|(\lambda + \tilde{A}(0))\bar{u}\|_{\mathrm{L}^p(0,T';\mathrm{L}^q_\sigma(\Omega_0))}\Big) \\
&\leq c\Big(\|\bar{u}_t\|_{\mathrm{L}^p(0,T';\mathrm{L}^q_\sigma(\Omega_0))} + \|(\lambda + A(\cdot))\bar{u}\|_{\mathrm{L}^p(0,T';\mathrm{L}^q_\sigma(\Omega_0))}\Big) \\
&\leq c\Big(\|L(\cdot)\bar{u}\|_{\mathrm{L}^p(0,T';\mathrm{L}^q_\sigma(\Omega_0))} + \|u_0\|_{\mathscr{F}^p_{1+\tilde{A}(0)}}\Big),
\end{aligned}
$$

where we used the resolvent estimate for the Stokes operator (1.18), the norm equivalence proven in Lemma 2.3.3, and the maximal L^p-regularity of the non-autonomous shifted problem.

Let us now prove that the operator $\mathcal{F}_{\bar{u}}$ fulfils the assumptions of Lemma 3.1.1. Using Lemma 3.2.1 we get that

$$\|\mathcal{F}_{\bar{u}}(g)\|_{\mathrm{L}^p(0,T';\mathrm{L}^q_\sigma(\Omega_0))} \leq c\|L(\cdot)^{-1}g + \bar{u}\|^2_{W^p_{\tilde{A}(0)}}.$$

So we can conclude that

$$\|L(\cdot)^{-1}g + \bar{u}\|_{W^p_{\tilde{A}(0)}} \leq c\Big(\|g\|_{\mathrm{L}^p(0,T';\mathrm{L}^q_\sigma(\Omega_0))} + \|L(\cdot)\bar{u}\|_{\mathrm{L}^p(0,T';\mathrm{L}^q_\sigma(\Omega_0))} + \|u_0\|_{\mathscr{F}^p_{1+\tilde{A}(0)}}\Big).$$

With the same ideas we obtain for $g_1, g_2 \in \mathrm{L}^p(0,T';\mathrm{L}^q_\sigma(\Omega_0))$ the estimate

$$
\begin{aligned}
&\|\mathcal{F}_{\bar{u}}(g_1) - \mathcal{F}_{\bar{u}}(g_2)\| \\
&\leq c\Big(\big\|L(\cdot)^{-1}g_1 - L(\cdot)^{-1}g_2\big\|_{W^p_{\tilde{A}(0)}} \big\|L(\cdot)^{-1}g_1 + \bar{u}\big\|_{W^p_{\tilde{A}(0)}} \\
&\quad + \big\|L(\cdot)^{-1}g_2 + \bar{u}\big\|_{W^p_{\tilde{A}(0)}} \big\|L(\cdot)^{-1}g_1 - L(\cdot)^{-1}g_2\big\|_{W^p_{\tilde{A}(0)}}\Big) \\
&\leq c\Big(\|g_1 - g_2\|\big(2\big(\|\bar{u}\|_{W^p_{\tilde{A}(0)}} + \|u_0\|_{\mathscr{F}^p_{1+\tilde{A}(0)}}\big) + \|g_1\| + \|g_2\|\big)\Big).
\end{aligned}
$$

Here, we denote $\| \cdot \| := \| \cdot \|_{L^p(0,T';L^q_\sigma(\Omega_0))}$. Hence we can apply Lemma 3.1.1 provided that $\|\overline{u}\|$ and $\|u_0\|$ are sufficiently small, which is fulfilled if T' is small enough. Finally, Lemma 3.2.2 implies that the fixed point of $\mathcal{F}_{\overline{u}}$ yields a solution via $u = L(\cdot)^{-1}g + \overline{u}$.

Let us finally prove the uniqueness of the solution to (3.2). Therefore, let u_1 with corresponding fixed point g_1 denote the solution constructed above and let u_2 with g_2 denote another solution. Furthermore, Lemma 3.1.1 yields the existence of a constant $r > 0$ such that the fixed point of $\mathcal{F}_{\overline{u}}$ is unique in the class of solutions with norm less or equal then r.

By choosing $T'' \leq T'$ sufficiently small, we obtain that

$$\|g_2\|_{L^p(0,T'';L^q_\sigma(\Omega_0))} < r$$

and thus since g_1 and g_2 are fixed points to the operator $\mathcal{F}_{\overline{u}}$, where we restrict ourselves to the interval $[0,T'']$, we obtain $g_1 = g_2$ on the interval $[0,T'')$, and hence this also holds for u_1 and u_2. Let

$$\tilde{T} := \begin{cases} T', & \text{if } \|g_1\|_{L^p(0,T';L^q_\sigma(\Omega_0))} < r, \\ \inf\{T'' \mid \|g_1\|_{L^p(0,T'';L^q_\sigma(\Omega_0))} = r\}, & \text{else.} \end{cases}$$

Then by iterating the argument above we obtain $g_1 = g_2$ on $[0,\tilde{T})$ and since a strong solution is continuous with values in $L^q_\sigma(\Omega_0)$ we obtain $u_1 = u_2$ on $[0,\tilde{T}]$. If $\tilde{T} = T'$, the proof is already complete.

Otherwise, let us consider the Navier–Stokes equation in $[\tilde{T},T']$ with initial value $u_1(\tilde{T}) = u_2(\tilde{T})$ and with the same arguments as above we obtain that there exists an $\varepsilon > 0$ such that $u_1 = u_2$ on $[\tilde{T},\tilde{T}+\varepsilon)$. Again with the continuity these solutions also coincide in $[\tilde{T},\tilde{T}+\varepsilon]$. Thus, the interval

$$\Lambda := \{t \in [0,T') \mid u_1(\tau) = u_2(\tau) \text{ for all } \tau \leq t\}$$

is relatively open and closed, hence $\Lambda = [0,T')$. $\qquad\square$

As the last and most important result of this chapter let us prove the existence result to solutions to the Navier–Stokes equations in $[0,\infty)$ with small initial data.

Theorem 3.2.4. *Let $1 < q < \frac{3}{2}$, $1 < p < \infty$ such that $\frac{2}{p} + \frac{3}{q} = 3$, and let ψ and ϕ fulfil Assumption 2.0.1. Let $u_0 \in \mathscr{J}^p_{A(0)}$ and let us assume* that $f \in \mathrm{L}^p(0, \infty; \mathrm{L}^q_\sigma(\Omega_0))$. *Then there exists a $\varepsilon > 0$ with the following property: Assume that*

$$\|u_0\|_{\mathscr{I}^p_{A(0)}} + \|f\|_{\mathrm{L}^p(0,\infty;\mathrm{L}^q_\sigma(\Omega_0))} + \|\phi - Id\|_{C^{3,1}_b} < \varepsilon.$$

Then system (3.2) *possesses a unique strong solution u.*

Proof. Note that if ε is sufficiently small, we obtain by (2.20) that the family of norms

$$\left(\|\tilde{A}(t)u\|_q\right)_{t\in[0,\infty)}$$

for $u \in \mathcal{D}(A)$ is equivalent and due to Lemma 2.1.7 and Lemma 2.4.1 we obtain that

$$\left(\|A(t)u\|_q\right)_{t\in[0,\infty)}$$

also defines a family of equivalent norms. Hence we can copy the proof of the Theorem above to obtain the desired result. $\qquad\square$

CHAPTER 4

Decay of a Weak Solution to the Navier–Stokes Equations

In this fourth chapter we will turn to a slightly different problem. As discussed in the preface, we would like to prove a decay result for weak solutions to the Navier–Stokes equations with non-homogeneous boundary data. Therefore, let us consider the equations

$$
\begin{aligned}
u_t - \Delta u + u \cdot \nabla u + \nabla p &= f, \\
\operatorname{div} u &= 0, \\
u_{|\partial \Omega} &= \beta, \\
u(0) &= u_0
\end{aligned}
$$

(4.1)

in a domain $\Omega \subset \mathbb{R}^3$ with compact $C^{1,1}$ boundary, *i.e.*, in a bounded domain or in an exterior domain. The aim of this chapter is to prove the existence of a weak solution to (4.1), which decays exponentially, if Ω is bounded, or polynomially, if Ω is an exterior domain.

This chapter is organized as follows. The existence of weak solutions to (4.1) is discussed in Section 4.1. The decay result for a weak solution if Ω is a bounded domain will be discussed in Section 4.2 and finally, the main result about the decay in the exterior domain case is proved in Section 4.3. Note that these results of this chapter are published in [17].

4.1 Existence of Weak Solutions to the Navier–Stokes Equations

Let us start by defining the notion of a weak solution to (4.1). These equations contain at least two terms which might be difficult to handle. One is the non-homogeneous boundary data, the second the non-linear term. As usual, we will split this into two different problems and we will start with the non-homogeneous boundary data. Therefore, let us consider the *stationary Stokes equations*

$$(4.2) \qquad -\Delta b + \nabla p = 0, \quad \operatorname{div} b = 0, \quad b_{|\partial\Omega} = \beta.$$

To define solutions to (4.2), let us introduce the space

$$C_{0,\sigma}^2(\overline{\Omega}) := \big\{ v \in C^2(\overline{\Omega}) \mid \operatorname{div} v = 0,$$
$$\operatorname{supp}(v) \subset \overline{\Omega} \text{ is compact and } v = 0 \text{ on } \partial\Omega \big\}.$$

Definition 4.1.1. Let $\Omega \subset \mathbb{R}^3$ be a domain with compact $C^{1,1}$ boundary. Let $1 < q < \infty$ and let $\beta \in W^{-\frac{1}{q},q}(\partial\Omega)$. A vector field $b \in L^q(\Omega)$ is called *very weak solution* to (4.2), if

$$-\int_\Omega b \cdot \Delta\varphi \mathrm{d}x = -\int_{\partial\Omega} \beta \cdot \big(n \cdot \nabla\varphi\big)\mathrm{d}\sigma$$

holds for all $\varphi \in C_{0,\sigma}^2(\overline{\Omega})$, $\operatorname{div} b = 0$ in the sense of distributions, and $b_{|\partial\Omega} \cdot n = \beta \cdot n$. Here, $n = n(x)$ denotes the outer normal vector of Ω.

In [16, Theorem 1.6] it is proved that for every $\beta \in W^{-\frac{1}{q},q}(\partial\Omega)$ there exists a unique very weak solution to (4.2), and the solution operator is continuous. Furthermore, since the space of test functions does not depend on q, we obtain that if $\beta \in W^{-\frac{1}{q},q}(\partial\Omega) \cap W^{-\frac{1}{p},p}(\partial\Omega)$ for some $p \neq q$, then the L^p-solution coincides with the L^q-solution.

Next, let us consider the instationary Stokes equations, *i.e.*,

$$(4.3) \qquad \begin{aligned} b_t - \Delta b + \nabla p &= 0, \quad \operatorname{div} b = 0, \\ b_{|\partial\Omega} &= \beta, \quad b(0) = 0. \end{aligned}$$

Definition 4.1.2. Let $\Omega \subset \mathbb{R}^3$ be a domain with compact $C^{1,1}$ boundary and let $1 < q < \infty$, $1 < s < \infty$ and $0 < T \leq \infty$. Let $\beta \in L^s(0, T; W^{-\frac{1}{q}, q}(\partial\Omega))$. A function $b \in L^s(0, T; L^q(\Omega))$ is called *very weak solution* to (4.3) if

$$-\langle b, \varphi_t \rangle_{\Omega, T} - \langle b, \Delta\varphi \rangle_{\Omega, T} = -\langle \beta, n \cdot \nabla\varphi \rangle_{\partial\Omega, T}$$

holds true for all $\varphi \in C_0^1([0, T); C_{0,\sigma}^2(\overline{\Omega}))$, and $\operatorname{div} b = 0$ in the sense of distributions.

Note that the instationary Stokes equations with non-homogeneous boundary data is also uniquely solvable and we have the representation formula

$$(4.4) \qquad b(t) = A_q \int_0^t e^{-(t-\tau)A_q} \gamma(\tau) \mathrm{d}\tau,$$

where $\gamma(t)$ denotes the unique very weak solution to the stationary Stokes equations with boundary data $\beta(t)$, see [16, Lemma 4.1]. Furthermore, the solution formula implies that the solution operator is continuous, and that if

$$\beta \in L^{s_1}(0, T; W^{-\frac{1}{q_1}, q_1}(\partial\Omega)) \cap L^{s_2}(0, T; W^{-\frac{1}{q_2}, q_2}(\partial\Omega)),$$

then the corresponding solutions coincide.

Let formally (u, p) denote a solution to (4.1) and (b, p') a solution to (4.3). Then $v := u - b$ and $\tilde{p} := p - p'$ is a solution to the *perturbed Navier–Stokes system*

$$(4.5) \qquad \begin{aligned} v_t - \Delta v + (v+b) \cdot \nabla(v+b) + \nabla\tilde{p} &= f, \quad \operatorname{div} v = 0, \\ v_{|\partial\Omega} &= 0, \quad v(0) = u_0. \end{aligned}$$

Since v and b are solenoidal, we can rewrite the non-linear term as

$$(v+b) \cdot \nabla(v+b) = \operatorname{div}\Big((v+b) \otimes (v+b)\Big),$$

where $\otimes$ denotes the dyadic product and the divergence is taken column-wise. Note that due to this representation we do not need any integrability conditions of the derivative of b in a weak formulation. Furthermore, since the solvability of (4.3) is well understood, we will assume that b and not β is part of the data.

Definition 4.1.3. Let $\Omega \subset \mathbb{R}^3$ be a domain with compact $C^{1,1}$ boundary and $0 < T \leq \infty$. Let $s_0 \in (2, \infty)$ and $q_0 \in (3, \infty)$ such that

$$\frac{2}{s_0} + \frac{3}{q_0} = 1$$

and assume that

$$b \in \mathrm{L}^4(0, T; \mathrm{L}^4(\Omega)) \cap \mathrm{L}^{s_0}(0, T; \mathrm{L}^{q_0}(\Omega)),$$

$f = \operatorname{div} F$ with $F \in \mathrm{L}^2(0, T; \mathrm{L}^2(\Omega))$, and $u_0 \in \mathrm{L}^2_\sigma(\Omega)$. Then a vector field v and $(0, T) \times \Omega$ is called a *Leray–Hopf type weak solution* to (4.5) if the following conditions are satisfied:

 (i) We have $v \in \mathrm{L}^\infty(0, T; \mathrm{L}^2_\sigma(\Omega))$ and $v \in \mathrm{L}^2(0, T; \dot{W}_0^{1,2}(\Omega))$.

 (ii) For each test function $\varphi \in C_0^\infty([0, T); C_{0,\sigma}^\infty(\Omega))$ the equality

$$
(4.6) \quad
\begin{aligned}
&- \langle v, \varphi_t \rangle_{\Omega,T} + \langle \nabla v, \nabla \varphi \rangle_{\Omega,T} - \Big\langle (v + b) \otimes (v + b), \nabla \varphi \Big\rangle_{\Omega,T} \\
&= \langle u_0, \varphi(0) \rangle_\Omega - \langle F, \nabla \varphi \rangle_{\Omega,T}
\end{aligned}
$$

is fulfilled.

 (iii) The *energy inequality*

$$
(4.7) \quad
\begin{aligned}
\frac{1}{2}\|v(t)\|_2^2 + \int_0^t \|\nabla v\|_2^2 \mathrm{d}\tau \\
\leq \frac{1}{2}\|u_0\|_2^2 - \int_0^t \langle F - (v + b) \otimes b, \nabla v \rangle \mathrm{d}\tau
\end{aligned}
$$

holds for all $t \in (0, T)$.

Note that every weak solution in the sense of Definition 4.1.3 can be redefined on a nullset such that the redefined function is a weak solution and weakly continuous with values in $\mathrm{L}^2_\sigma(\Omega)$. We will always assume that a weak solution to (4.1) is weakly continuous.

The solvability of the perturbed Navier–Stokes system was recently studied by several authors, see for instance [2, 14, 15, 44]. As in many cases, the solution is constructed using an approximation process and since this approximation is important in the proof of the decay result, we will shortly present the idea of the proof.

For $k \in \mathbb{N}$ and $1 < q < \infty$ let

$$J_k := \left(1 + \frac{1}{k}A^{\frac{1}{2}}_q\right)^{-1}$$

denote the *Yosida approximation*. Since A_q and hence its root are sectorial operators, we obtain that $(J_k)_k$ and $(\frac{1}{k}A^{\frac{1}{2}}_q J_k)_k$ are well defined and uniformly bounded operators in $L^q_\sigma(\Omega)$, and $J_k v \to v$ for all $v \in L^q_\sigma(\Omega)$ as $k \to \infty$.

Then we may consider the *approximate perturbed Navier–Stokes system*

$$(4.8) \qquad \begin{aligned} v_t - \Delta v + (J_k v + b) \cdot \nabla(v + b) + \nabla p &= f, \quad \operatorname{div} v = 0, \\ v_{|\partial\Omega} &= 0, \quad v(0) = u_0. \end{aligned}$$

This system possesses a unique weak solution $v = v_k$ in the Leray–Hopf class such that the representation formula

$$(4.9) \qquad v_k(t) = e^{-tA} u_0 + \int_0^t A^{\frac{1}{2}} e^{-(t-\tau)A} (A^{-\frac{1}{2}} P \operatorname{div}) \hat{F}(\tau) \mathrm{d}\tau$$

holds. Here, we define

$$\hat{F} := F - (J_k v_k + b) \otimes (v_k + b),$$

and the solution fulfils the differential equation

$$(4.10) \qquad \frac{1}{2}\partial_t \|v_k(t)\|_2^2 + \|\nabla v_k(t)\|_2^2 = -\langle F - (J_k v + b) \otimes b, \nabla v_k(t)\rangle$$

in the sense of distributions.

Due to the uniform boundedness of the approximate solutions $(v_k)_k$ in the Leray–Hopf class and a compactness argument we obtain the existence of a function v in the Leray–Hopf class and of a subsequence (v_{k_j}) such that for any bounded subdomain $\Omega' \subset \Omega$ we have

$$(4.11) \qquad \begin{aligned} v_{k_j} &\overset{*}{\rightharpoonup} v && \text{in} \quad L^\infty(0,\infty; L^2(\Omega)), \\ \nabla v_{k_j} &\rightharpoonup \nabla v && \text{in} \quad L^2(0,\infty; L^2(\Omega)), \\ v_{k_j} &\to v && \text{in} \quad L^2(0,T'; L^2(\Omega')), && \text{for all } T' < \infty, \\ v_{k_j}(t) &\to v(t) && \text{in} \quad L^2(\Omega') && \text{for almost all } t \geq 0. \end{aligned}$$

As it is proved in [14], the vector field v is a Leray–Hopf type weak solution to (4.5). Let us prove as a last result of this section that the convergence result above implies that v has a suitable decay property provided that the approximation sequence $(v_k)_k$ fulfils the same.

Lemma 4.1.4. *Let $(v_k)_k$, $(v_{k_j})_j$ and v be as above.*

(i) *Assume that*

$$\limsup_{\substack{t \to \infty \\ k \in \mathbb{N}}} \|v_k(t)\|_2 = 0.$$

Then

$$\lim_{t \to \infty} \|v(t)\|_2 = 0.$$

(ii) *Assume that*

$$\|v_k(t)\| \le cf(t)$$

with $f(t) = e^{-\gamma t}$ or $f(t) = t^{-\gamma}$ and $c > 0$ is chosen uniformly in k. Then

$$\|v(t)\|_2 \in O(f(t)).$$

Proof. Let $t \in (0, \infty)$ be arbitrary and $\varepsilon > 0$ and – in case (ii) $-\varepsilon < cf(t)$. Then there exists a bounded subdomain $\Omega' \subset \Omega$ such that

$$\|v(t)\|_{L^2(\Omega)} \le \|v(t)\|_{L^2(\Omega')} + \varepsilon.$$

If $v_{k_j}(t)$ converges strongly in $L^2(\Omega')$ to $v(t)$, the result is obvious. Otherwise, choose a $\varphi \in L^2(\Omega')$, $\|\varphi\|_{L^2(\Omega')} = 1$, such that

$$\|v(t)\|_{L^2(\Omega)} \le \|v(t)\|_{L^2(\Omega')} + \varepsilon = \langle v(t), \varphi \rangle + \varepsilon,$$

and due to the weak continuity of v and the almost everywhere strong convergence we can choose a $\tau \in (0, \infty)$ such that $|t - \tau| < 1$, $v_{k_j}(\tau)$ converges strongly, and $|\langle v(t), \varphi \rangle - \langle v(\tau), \varphi \rangle| < \varepsilon$. Furthermore –in case (ii)– $|f(t) - f(\tau)| < c^{-1}\varepsilon$. Therefore, let us estimate further

$$\begin{aligned}
\|v(t)\|_{L^2(\Omega)} &\le \langle v(\tau), \varphi \rangle + 2\varepsilon \\
&= \lim_{j \to \infty} \langle v_{k_j}(\tau), \varphi \rangle + 2\varepsilon \\
&\le \sup_{k \in \mathbb{N}} \|v_k(\tau)\|_{L^2(\Omega)} + 2\varepsilon.
\end{aligned}$$

The proof of (i) is completed. In case (ii) note that

$$\|v(t)\|_{L^2(\Omega)} \le cf(\tau) + 2\varepsilon \le cf(t) + 3\varepsilon \le 4cf(t)$$

and we are done. $\qquad\square$

4.2 Decay of a Weak Solution to Navier–Stokes Equations in Bounded Domains

The aim of this section is to prove a decay result for weak solutions to the Navier–Stokes equations (4.5), see also [17].

Theorem 4.2.1. *Let $\Omega \subset \mathbb{R}^3$ be a bounded domain with $\partial\Omega \in C^{1,1}$. Let $f = \operatorname{div} F$, $F \in L^2(0, \infty; L^2(\Omega))$, $u_0 \in L^2_\sigma(\Omega)$, and*

$$b \in L^4(0, T; L^4(\Omega)) \cap L^{s_0}(0, T; L^{q_0}(\Omega)), \quad \frac{2}{s_0} + \frac{3}{q_0} = 1.$$

Then there exists a weak solution v to (4.5) such that

$$\lim_{t \to \infty} \|v(t)\|_2 = 0.$$

Let $\alpha \in (0, 1)$, $\beta > 0$, and assume in addition that

(4.12) $\quad \|F\|^2_{L^2(\alpha t, t; L^2(\Omega))} + \|b\|^4_{L^4(\alpha t, t; L^4(\Omega))} + \|b\|^{s_0}_{L^{s_0}(\alpha t, t; L^{q_0}(\Omega))} \in O(e^{-\beta t})$

as $t \to \infty$. Then there exists a weak solution v to (4.5) such that

$$\|v(t)\|_2 \in O(\exp(-t\gamma)) \ as \ t \to \infty.$$

Here, $\gamma := \frac{1}{2} \min\{(1 - \alpha)\rho, \beta\}$ and $\sqrt[-2]{\rho}$ denotes the largest constant, such that Poincaré's inequality holds, i.e., $\rho\|u\|^2_2 \le \|\nabla u\|^2_2$ for all $u \in W^{1,2}_0(\Omega)$.

Proof. We will show that the statements hold for the sequence of solutions $(v_k)_{k \in \mathbb{N}}$ of the approximate perturbed Navier–Stokes system (4.8), where the appearing constants do *not* depend on k, and Lemma 4.1.4 will imply this result for the weak solution v.

From now on we will omit the index k of v_k. Due to the energy equality for $v = v_k$ and the interpolation estimate

$$\|v\|_{(\frac{1}{2} - \frac{1}{q_0})^{-1}} \le c\|v\|^{2/s_0}_2 \|\nabla v\|^{1 - 2/s_0}_2$$

which is valid due to

$$\|v\|_{(\frac{1}{2}-\frac{1}{q_0})^{-1}} \le \|v\|_2^{\alpha'}\|v\|_6^{1-\alpha'} \le c\|v\|_2^{\alpha'}\|\nabla v\|_2^{1-\alpha'}$$

and α' is choosen such that

$$\frac{1}{2} - \frac{1}{q_0} = \frac{\alpha'}{2} + \frac{1-\alpha'}{6} = \frac{1}{3}\alpha' + \frac{1}{6}$$

or equivalently

$$\alpha' = 1 - \frac{3}{q_0} = \frac{2}{s_0},$$

we get from Hölder's and Young's inequality using (4.10) that

$$
\begin{aligned}
\frac{1}{2}\partial_t\|v(t)\|_2^2 + \|\nabla v(t)\|_2^2 &= -\Big(F - (J_k v + b) \otimes b, \nabla v\Big)\\
&\le \|F\|_2\|\nabla v\|_2 + \|b\|_4^2\|\nabla v\|_2 + \|v\|_{(\frac{1}{2}-\frac{1}{q_0})^{-1}}\|b\|_{q_0}\|\nabla v\|_2\\
&\le \|F\|_2\|\nabla v\|_2 + \|b\|_4^2\|\nabla v\|_2 + \|v\|_{2,\infty}^{2/s_0}\|\nabla v\|_2^{2-2/s_0}\|b\|_{q_0}\\
&\le \frac{1}{2}\|\nabla v(t)\|_2^2 + 2\|F\|_2^2 + 2\|b\|_4^4 + c\|b\|_{q_0}^{s_0};
\end{aligned}
$$

here $c > 0$ is a constant depending on a uniform bound of the energies $\|v\|_{2,\infty}^2$. Absorbing the term $\frac{1}{2}\|\nabla v(t)\|_2^2$ from the right-hand side we arrive at the estimate

$$(4.13) \qquad \partial_t\|v(t)\|_2^2 + \|\nabla v(t)\|_2^2 \le 4\|b(t)\|_4^4 + 4\|F(t)\|_2^2 + c\|b(t)\|_{q_0}^{s_0}.$$

So far, we did not use the fact that Ω is bounded. Since Ω is bounded we exploit Poincaré's inequality $\rho\|v\|_2^2 \le \|\nabla v\|_2^2$ and conclude from (4.13) that

$$(4.14) \qquad \partial_t\|v(t)\|_2^2 + \rho\|v(t)\|_2^2 \le c\left(\|F(t)\|_2^2 + \|b(t)\|_{q_0}^{s_0} + \|b(t)\|_4^4\right).$$

Multiplying (4.14) with $e^{\rho t}$ and integrating from 0 to t leads for $\alpha \in (0,1)$ to the estimate

(4.15)

$$
\begin{aligned}
e^{\rho t}\|v(t)\|_2^2 - \|u_0\|_2^2 &\le c\int_0^t e^{\rho\tau}\left(\|F(\tau)\|_2^2 + \|b(\tau)\|_{q_0}^{s_0} + \|b(\tau)\|_4^4\right)\mathrm{d}\tau\\
&\le ce^{\alpha\rho t}\left(\|F\|_{2,2}^2 + \|b\|_{q_0,s_0}^{s_0} + \|b\|_{4,4}^4\right)\\
&\quad + ce^{\rho t}\left(\|F\|_{L^2(\alpha t,t;L^2)}^2 + \|b\|_{L^{s_0}(\alpha t,t;L^{q_0})}^{s_0} + \|b\|_{L^4(\alpha t,t;L^4)}^4\right).
\end{aligned}
$$

Note that by the assumptions on F and b

$$\|F\|^2_{L^2(\alpha t,t;L^2(\Omega))} + \|b\|^{s_0}_{L^{s_0}(\alpha t,t;L^{q_0}(\Omega))} + \|b\|^4_{L^4(\alpha t,t;L^4(\Omega))} \to 0 \quad \text{as } t \to \infty.$$

Finally, dividing (4.15) by $e^{\rho t}$ we conclude that $\|v(t)\|_2 \to 0$ as $t \to \infty$. Moreover, under the assumption (4.12) the statement on the exponential decay follows immediately. $\qquad\qquad\square$

4.3 Decay of a Weak Solution to Navier–Stokes Equations in Exterior Domains

Let us consider the main statements of this chapter, *i.e.*, a decay result for a suitable weak solution if Ω is an exterior domain. Let us start with an elementary Lemma.

Lemma 4.3.1. *Let* $m \geq 1$.

(i) *Let* $\psi \in L^\infty(0,\infty)$ *satisfy* $\lim_{s\to\infty} \psi(s) = 0$. *Then*

$$\frac{1}{(t/2)^m} \int_{\frac{t}{2}}^{t} (s - t/2)^{m-1} \psi(s)\,ds \to 0 \text{ as } t \to \infty.$$

Here, we define

$$\lim_{s\to\infty} \psi(s) = 0 \quad \textit{iff} \quad \lim_{s\to\infty} \|\psi\|_{L^\infty(s,\infty)} = 0.$$

(ii) *Let* $\psi \in L^1(0,\infty)$. *Then*

$$\frac{1}{(t/2)^m} \int_{\frac{t}{2}}^{t} (s - t/2)^{m} \psi(s)\,ds \to 0 \text{ as } t \to \infty.$$

(iii) *Let* $\phi \in L^2(0,\infty)$ *and*

$$\psi(s) := \int_0^s (s - \tau)^{\frac{1}{4} - \frac{3}{2q}} \phi(\tau)\,d\tau,$$

where $\frac{6}{5} < q < 2$. *Then*

$$\frac{1}{(t/2)^m} \int_{\frac{t}{2}}^{t} (s - t/2)^{m-1} \psi^2(s)\,ds \leq c t^{\frac{3}{2} - \frac{3}{q}} \|\phi\|^2_{L^2(0,\infty)}$$

with some constant c which does not depend on ϕ.

(iv) *Let* $\phi \in L^r(0, \infty)$, $1 \leq r \leq \infty$, $\phi \geq 0$, *and*

$$\psi(s) := \int_0^s (s - \tau)^{\frac{1}{4} - \frac{3}{2q}} \phi(\tau) d\tau,$$

where $\frac{6}{5} < q$. *Then*

$$\frac{1}{(t/2)^m} \int_{\frac{t}{2}}^t (s - t/2)^{m-1} \psi(s) ds \leq ct^{\frac{5}{4} - \frac{3}{2q} - \frac{1}{r}} \|\phi\|_{L^r(0,\infty)}$$

with some absolute constant $c > 0$.

Proof. The statement (i) is trivial and (ii) is a consequence of Lebesgue's Theorem of Dominated Convergence. To prove (iii) we obtain by Hölder's inequality

$$\psi(s)^2 \leq \int_0^s (s - \tau)^{\frac{1}{4} - \frac{3}{2q}} d\tau \int_0^s (s - \tau)^{\frac{1}{4} - \frac{3}{2q}} \phi^2(\tau) d\tau$$

$$\leq cs^{\frac{5}{4} - \frac{3}{2q}} \int_0^s (s - \tau)^{\frac{1}{4} - \frac{3}{2q}} \phi^2(\tau) d\tau$$

for some absolute constant c. Hence we obtain by Tonelli's Theorem

$$\frac{1}{(t/2)^m} \int_{\frac{t}{2}}^t (s - t/2)^{m-1} \psi^2(s) ds \leq \frac{c}{t} \int_{\frac{t}{2}}^t \int_0^s s^{\frac{5}{4} - \frac{3}{2q}} (s - \tau)^{\frac{1}{4} - \frac{3}{2q}} \phi^2(\tau) d\tau ds$$

$$\leq ct^{\frac{1}{4} - \frac{3}{2q}} \int_0^t \phi^2(\tau) \Big(\int_\tau^t (s - \tau)^{\frac{1}{4} - \frac{3}{2q}} ds \Big) d\tau$$

$$\leq ct^{\frac{3}{2} - \frac{3}{q}} \|\phi\|_{L^2(0,\infty)}^2.$$

Finally, let us prove (iv). Let us first assume that $r > 1$. Then, note that the assumption on q implies $\frac{1}{4} - \frac{3}{2q} > -1$ and let $\frac{1}{4} - \frac{3}{2q} = \gamma + \gamma'$, γ and γ' are chosen such that

$$r\gamma > -1, \quad r'\gamma' > -1,$$

and r' denote the Hölder conjugate exponent to r. This is possible since if $\tilde{\gamma} = -\frac{1}{r}$, and $\tilde{\gamma}' = -1 - \tilde{\gamma}$, then we obtain

$$\tilde{\gamma}'r' = (-1 - \tilde{\gamma})\frac{r}{r - 1} = \frac{-r + 1}{r - 1} = -1;$$

hence the strict inequality $\frac{1}{4} - \frac{3}{2q} > -1$ ensures us to choose γ and γ' suitably. Then we obtain using Hölder's inequality and Tonelli's Theorem

$$\frac{1}{(t/2)^m} \int_{\frac{t}{2}}^{t} (s - t/2)^{m-1} \psi(s) ds$$

$$= ct^{-m} \int_{\frac{t}{2}}^{t} (s - t/2)^{m-1} \int_0^s (s - \tau)^{\frac{1}{4} - \frac{3}{2q}} \phi(\tau) d\tau ds$$

$$\leq ct^{-m} \int_{\frac{t}{2}}^{t} (s - t/2)^{m-1} \|(s - \cdot)^{\gamma'}\|_{L^{r'}(0,s)} \|(s - \cdot)^{\gamma} \phi\|_{L^r(0,s)} ds$$

$$\leq ct^{-1} \Big(\int_{\frac{t}{2}}^{t} \int_0^s (s - \tau)^{\gamma' r'} d\tau ds \Big)^{\frac{1}{r'}} \Big(\int_{\frac{t}{2}}^{t} \int_0^s \phi(\tau)^r (s - \tau)^{\gamma r} d\tau ds \Big)^{\frac{1}{r}}$$

$$\leq ct^{-1 + \gamma' + \frac{2}{r'}} \Big(\int_0^t \phi(\tau)^r \int_\tau^t (s - \tau)^{\gamma r} ds d\tau \Big)^{\frac{1}{r}}$$

$$\leq ct^{-1 + \gamma' + \frac{2}{r'} + \gamma + \frac{1}{r}} \|\phi\|_{L^r}$$

$$\leq ct^{\frac{5}{4} - \frac{3}{2q} - \frac{1}{r}} \|\phi\|_{L^r}.$$

In case $r = 1$, one can choose $\gamma = \frac{1}{4} - \frac{3}{2q}$ and skip the first use of Hölder's inequality. $\qquad \square$

In the final main theorem about the decay of a suitable weak solution to (4.5) in exterior domains, we will prove that assuming the data tends to zero fast enough, then the L^2-norm of the solution will tend to zero almost as fast as $t^{-3/4}$.

Theorem 4.3.2. *Let $\Omega \subset \mathbb{R}^3$ be an exterior domain with $\partial\Omega \in C^{1,1}$. Let $f = \operatorname{div} F$, $F \in L^2(0, \infty; L^2(\Omega))$, let $u_0 \in L_\sigma^2(\Omega)$ and let*

$$b \in L^4(0, \infty; L^4(\Omega)) \cap L^{s_0}(0, \infty; L^{q_0}(\Omega)), \quad \frac{2}{s_0} + \frac{3}{q_0} = 1.$$

Assume that

$$\nabla b \in L^{s_1}(0, \infty; L^{q_1}(\Omega)), \quad \frac{2}{s_1} + \frac{3}{q_1} = 2, \ q_1 > 3,$$

$$b \in L^{s_2}(0, \infty; L^{q_2}(\Omega)), \quad \frac{2}{s_2} + \frac{3}{q_2} =: S(b), \ 3 < q_2 \leq 4, \ s_2 \geq 2,$$

$$b \cdot \nabla b \in L^1(0, \infty; L^2(\Omega)),$$

$$\nabla F \in L^1(0, \infty; L^2(\Omega)),$$

$$F \in L^{s_3}(0, \infty; L^{q_3}(\Omega)), \quad \frac{2}{s_3} + \frac{3}{q_3} =: S(F), \ \frac{3}{2} < q_3 \leq 2, \ s_3 \geq 1,$$

and that

$$\|\nabla F\|_{L^1(t/2,t,L^2(\Omega))} + \|\nabla b\|_{L^{s_1}(t/2,t,L^{q_1}(\Omega))} + \|b \cdot \nabla b\|_{L^1(t/2,t,L^2(\Omega))} \in O(t^{-\alpha_2})$$

as $t \to \infty$ *for some* $\alpha_2 > 0$*. Moreover, suppose that*

$$\|e^{-tA}u_0\|_2 \in O(t^{-\alpha_1})$$

for some $\alpha_1 \geq 0$*. Then there exists a weak solution* v *to* (4.5) *such that*

$$\|v(t)\|_2 \in O(t^{-\alpha}) \text{ as } t \to \infty,$$

where

$$\alpha := \min\left\{S(b) - \frac{5}{4}, \frac{S(F)}{2} - \frac{5}{4}, \alpha_1, \alpha_2\right\}.$$

Note that the restrictions on q_2 and q_3 imply that $\alpha < \frac{3}{4}$. Furthermore, if $f = 0$ and $\beta = 0$ and thus $b = 0$, Theorem 4.3.2 covers the results of W. Bochers and T. Miyakawa [5] except that we cannot achieve the borderline case $\alpha = \frac{3}{4}$.

Proof. As in the proof of Theorem 4.2.1, we will prove that the sequence of solutions $(v_k)_k$ to (4.8) has the desired decay property and Lemma 4.1.4 will imply that also the weak solution v to (4.5) has the desired decay property. For simplicity, we will omit the index k as far as possible.

Let us start with (4.10), hence we have

$$\frac{1}{2}\partial_t\|v(t)\|_2^2 \leq \frac{1}{2}\partial_t\|v(t)\|_2^2 + \|\nabla v(t)\|_2^2$$
$$= -\langle F - (J_k v + b) \otimes b, \nabla v(t)\rangle,$$

and the regularity assumptions on F and b ensure that we can integrate by parts to obtain

$$\frac{1}{2}\partial_t\|v(t)\|_2^2 \leq \langle \operatorname{div} F - (J_k v + b) \cdot \nabla b, v(t)\rangle$$
$$\leq \left(\|\nabla F\|_2 + \|J_k v \cdot \nabla b\|_2 + \|b \cdot \nabla b\|_2\right)\|v(t)\|_2$$

Using $\frac{1}{2}\partial_t\|v(t)\|_2^2 = \|v(t)\|_2\partial_t\|v(t)\|_2$, dividing by $\|v(t)\|_2$, and adding $\rho\|v(t)\|_2$ on both sides of the inequality we obtain

$$\partial_t\|v(\tau)\|_2 + \rho\|v(\tau)\|_2 \leq \|\nabla F\|_2 + \|J_k v \cdot b\|_2 + \|b \cdot \nabla b\|_2 + \rho\|v(\tau)\|_2.$$

Replacing ρ by $\rho(\tau) = m(\tau - t/2)^{-1}$ for some $m \geq 1$, multiplying the inequality above by $(\tau - t/2)^m$ and integrating τ from $t/2$ to t yields

(4.16)
$$\|v(t)\|_2 \leq \|\nabla F\|_{\mathrm{L}^1(t/2,t;\mathrm{L}^2(\Omega))} + \|J_k v \cdot \nabla b\|_{\mathrm{L}^1(t/2,t;\mathrm{L}^2(\Omega))}$$
$$+ \|b \cdot \nabla b\|_{\mathrm{L}^1(t/2,t;\mathrm{L}^2(\Omega))} + (t/2)^{-m} \int_{\frac{t}{2}}^{t} m(\tau - t/2)^{m-1}\|v(\tau)\|_2 \mathrm{d}\tau.$$

The assumptions immediately imply that the first and the third addend on the right hand side of (4.16) can be estimated by $ct^{-\alpha}$. To treat the second term, note that

$$J_k v \in \mathrm{L}^2(0,\infty;\mathrm{L}^6(\Omega)) \cap \mathrm{L}^\infty(0,\infty;\mathrm{L}^2(\Omega)),$$

and $(J_k v)_k$ is uniformly bounded in the spaces above. Thus, we have

$$J_k v \in \mathrm{L}^s(0,\infty;\mathrm{L}^q(\Omega))$$

for all $s \in (2,\infty)$ and $q \in (2,6)$ such that

$$\frac{2}{s} + \frac{3}{q} = \frac{3}{2}.$$

Let us choose q such that
$$\frac{1}{q} + \frac{1}{q_1} = \frac{1}{2}.$$
Thus we have for $s = s(q)$ that

$$\frac{1}{s} + \frac{1}{s_1} = \frac{1}{2}\left(\frac{3}{2} + 2 - \frac{3}{q} - \frac{3}{q_1}\right) = \frac{1}{2}\left(\frac{3}{2} + 2 - \frac{3}{2}\right) = 1$$

and hence we obtain for the second term on the right hand side of (4.16) that

$$\|J_k v \cdot \nabla b\|_{\mathrm{L}^1(t/2,t;\mathrm{L}^2(\Omega))} \leq \|v\|_{\mathrm{L}^s(0,\infty;\mathrm{L}^q(\Omega))}\|\nabla b\|_{\mathrm{L}^{s_1}(t/2,t;\mathrm{L}^{q_1}(\Omega))}.$$

This proves, that also this term decays like $t^{-\alpha}$.

Finally, we have to estimate the fourth addend in (4.16). Therefore, note that we have a representation formula for v by (4.9). Due to this representation formula, the fourth addend in (4.16) splits into six addends corresponding to the term including the initial value, which will be estimated first, and the five terms corresponding to $\hat{F}$ in (4.9) denoted by $T_0, ..., T_4$. Note that

$$(t/2)^{-m} \int_{\frac{t}{2}}^{t} m(\tau - t/2)^{m-1} \|e^{-\tau A}u_0\|_2 d\tau \le ct^{-\alpha}$$

is obvious by the assumption on the decay of $e^{-tA}u_0$. For the term T_0 corresponding to F we obtain

$$T_0 = \frac{m}{(t/2)^m} \int_{\frac{t}{2}}^{t} (s - t/2)^{m-1} \int_0^s \|A^{\frac{1}{2}} e^{-(s-\tau)A}(A^{-\frac{1}{2}}P\mathrm{div})F(\tau)\|_2 d\tau ds$$

$$\le \frac{c}{(t/2)^m} \int_{\frac{t}{2}}^{t} (s - t/2)^{m-1} \int_0^s |s - \tau|^{\frac{1}{4} - \frac{3}{2q_3}} \|(A^{-\frac{1}{2}}P\mathrm{div})F(\tau)\|_{q_3} d\tau ds$$

using (1.22) and (1.23), and next, the continuity of $A^{-\frac{1}{2}}P\mathrm{div}$ in L^{q_3} implies, see (1.21),

$$T_0 \le ct^{\frac{5}{4} - \frac{1}{2}S(F)} \|F\|_{\mathrm{L}^{s_3}(0,\infty;\mathrm{L}^{q_3}(\Omega))}$$

using Lemma 4.3.1. Note that we need to assume $q_3 > \frac{3}{2}$ since $A^{-\frac{1}{2}}P\mathrm{div}$ might be unbounded in $\mathrm{L}^{\frac{3}{2}}$. Obviously, the best possible decay rate can be proved with this method if q_3 is *close* to $\frac{3}{2}$ and $s_3 = 1$. In this case, T_0 decays like $t^{-\frac{3}{4} + \varepsilon}$.

The term T_4 corresponding to $b \otimes b$ can be estimated exactly in the same manner and we obtain

$$T_4 \le \frac{m}{(t/2)^m} \int_{\frac{t}{2}}^{t} (s - t/2)^{m-1} \int_0^s \|A^{\frac{1}{2}} e^{-(s-\tau)A}(A^{-\frac{1}{2}}P\mathrm{div})(b \otimes b)\|_2 d\tau ds$$

$$\le \frac{m}{(t/2)^m} \int_{\frac{t}{2}}^{t} (s - t/2)^{m-1} \int_0^s (s - \tau)^{\frac{1}{4} - \frac{3}{q_2}} \|b\|_{q_2}^2 d\tau ds$$

$$\le ct^{\frac{5}{4} - S(b)} \|b\|_{q_2, s_2}^2 .$$

Here, we used the boundedness of $A^{-\frac{1}{2}}P\mathrm{div}$ in $\mathrm{L}^{\frac{q_2}{2}}$ and hence the assumption $q_2 > 3$ is needed.

Let us consider the last three addends and therefore let

$$q \in (3/2, \min\{q_2/2, q_3\}) \subset (3/2, 2).$$

To estimate the term T_1 corresponding to $J_k v \otimes v$ let $\alpha' \in (0,1)$ be defined by

$$\frac{1}{2q} = \frac{\alpha'}{2} + \frac{1-\alpha'}{6},$$

and we estimate

$$
\begin{aligned}
T_1 &\leq \frac{m}{(t/2)^m} \int_{\frac{t}{2}}^{t} (s-t/2)^{m-1} \int_0^s \|A^{\frac{1}{2}} e^{-(s-\tau)A} (A^{-\frac{1}{2}} P\mathrm{div})(J_k v \otimes v)\|_2 d\tau ds \\
&\leq \frac{c}{(t/2)^m} \int_{\frac{t}{2}}^{t} (s-t/2)^{m-1} \int_0^s (s-\tau)^{\frac{1}{4}-\frac{3}{2q}} \|J_k v \otimes v\|_q d\tau ds \\
&\leq c t^{\frac{5}{4}-\frac{3}{2q}-1} \|J_k v \otimes v\|_{\mathrm{L}^1(0,t;\mathrm{L}^q(\Omega))}. \\
&\leq c t^{\frac{5}{4}-\frac{3}{2q}-1} \|v\|^{2\alpha'}_{\mathrm{L}^2(0,t;\mathrm{L}^2(\Omega))} \|v\|^{2-2\alpha'}_{\mathrm{L}^2(0,\infty;\mathrm{L}^6(\Omega))}
\end{aligned}
$$

Next, let us estimate the term corresponding to $b \otimes v$. We have

$$
\begin{aligned}
T_3 &\leq \frac{m}{(t/2)^m} \int_{\frac{t}{2}}^{t} (s-t/2)^{m-1} \int_0^s \|A^{\frac{1}{2}} e^{-(s-\tau)A} (A^{-\frac{1}{2}} P\mathrm{div})(b \otimes v)\|_2 d\tau ds \\
&\leq \frac{c}{(t/2)^m} \int_{\frac{t}{2}}^{t} (s-t/2)^{m-1} \int_0^s |s-\tau|^{\frac{1}{4}-\frac{3}{2q}} \|(A^{-\frac{1}{2}} P\mathrm{div})(b \otimes v)(\tau)\|_q d\tau ds.
\end{aligned}
$$

Choose s' such that $\frac{1}{s'} = \frac{1}{2} + \frac{1}{s_2}$ and we obtain

$$
\begin{aligned}
T_3 &\leq c t^{\frac{5}{4}-\frac{3}{2q}-\frac{1}{s'}} \|b \otimes v\|_{q,s'} \\
&\leq c t^{\frac{3}{4}-\frac{3}{2q}-\frac{1}{s_2}} \|b\|_{q_2,s_2} \|v\|^{\alpha''}_{\mathrm{L}^2(0,t;\mathrm{L}^2(\Omega))} \|v\|^{1-\alpha''}_{6,2},
\end{aligned}
$$

where α'' is chosen such that

$$\frac{1}{q} = \frac{1}{q_2} + \frac{\alpha''}{2} + \frac{1-\alpha''}{6}.$$

Obviously, T_2 can be estimated in the same way.

Summarising, due (4.16) and the estimates to the fourth terms in (4.16), we proved that

(4.17)
$$\|v(t)\|_2 \le ct^{-\alpha} + ct^{\frac{1}{4}-\frac{3}{2q}}\|v\|^{2\alpha'}_{L^2(0,t;L^2(\Omega))} + ct^{\frac{3}{4}-\frac{3}{2q}-\frac{1}{s_2}}\|v\|^{\alpha''}_{L^2(0,t;L^2(\Omega))}$$
$$\le ct^{-\alpha} + ct^{-\frac{1}{4}-\alpha'}\|v\|^{2\alpha'}_{L^2(0,t;L^2(\Omega))} + ct^{\frac{1}{2}-\frac{S(b)}{2}-\frac{\alpha''}{2}}\|v\|^{\alpha''}_{L^2(0,t;L^2(\Omega))}.$$

This will be the starting point of a boot strapping argument. Since $v \in L^\infty(0,\infty;L^2(\Omega))$, the growth rate of $\|v\|_{L^2(0,t;L^2(\Omega))}$ can be bounded. By inserting this in (4.17), we will obtain a first decay rate of $\|v(t)\|_{L^2(\Omega)}$ and this implies a stricter growth rate of $\|v\|_{L^2(0,t;L^2(\Omega))}$. Let us start with the details.

Let us assume first that $\alpha \le \frac{1}{4}$. Since $v \in L^\infty(0,\infty;L^2(\Omega))$ we obtain that
$$\|v\|_{L^2(0,t;L^2(\Omega))} \le ct^{\frac{1}{2}},$$

and thus we have

(4.18)
$$\|v(t)\|_2 \le ct^{-\alpha} + ct^{-\frac{1}{4}} + ct^{\frac{1}{2}-\frac{1}{2}S(b)}$$
$$\le ct^{-\alpha} + ct^{-\frac{1}{4}}.$$

To verify the last step in the estimate above note that $\alpha \le S(b) - \frac{5}{4}$ and hence if $\alpha \in (0,\frac{1}{4}]$ we obtain that $\frac{1}{2} - \frac{1}{2}S(b) \le -\alpha$.

In the next step let $\alpha \in (\frac{1}{4},\frac{3}{4q}]$. Then by (4.18) it is already proved that
$$\|v(t)\|_2 \le ct^{-\frac{1}{4}}$$

and hence
$$\|v\|_{L^2(0,t;L^2(\Omega))} \le ct^{\frac{1}{4}}.$$

Thus, we obtain by (4.17) that

(4.19)
$$\|v(t)\|_2 \le ct^{-\alpha} + ct^{-\frac{1}{4}-\frac{\alpha'}{2}} + ct^{\frac{1}{2}-\frac{S(b)}{2}-\frac{\alpha''}{4}}$$
$$\le ct^{-\alpha} + ct^{-\frac{3}{4q}} + ct^{\frac{1}{2}(\frac{5}{4}-S(b))-\frac{3}{4q}+\frac{3}{4q_2}}$$
$$\le ct^{-\alpha} + ct^{-\frac{3}{4q}} + ct^{-\frac{\alpha}{2}-\frac{3}{4q}+\frac{3}{4q_2}}$$
$$\le ct^{-\alpha} + ct^{-\frac{3}{4q}} + ct^{-\frac{\alpha}{2}-\frac{3}{4q}+\frac{3}{8q}}$$
$$\le ct^{-\alpha} + ct^{-\frac{3}{4q}} + ct^{-\frac{\alpha}{2}-\frac{3}{8q}}.$$

Here we used $q \leq \frac{q_2}{2}$. This proves the decay estimate provided that $\alpha \in [\frac{1}{4}, \frac{3}{4q})$ and since q can be chosen arbitrarily close to $\frac{3}{2}$, the statement is proved provided that $\alpha < \frac{1}{2}$. Let us prove next that we can get beyond $\frac{1}{2}$. Since $\|v(t)\|_2 \leq ct^{-\frac{3}{4q}}$ we obtain that

$$\|v\|_{L^2(0,t;L^2(\Omega))} \leq ct^{\frac{1}{2}-\frac{3}{4q}}$$

and hence by (4.17) we get

$$\|v(t)\|_2 \leq ct^{-\alpha} + ct^{-\frac{1}{4}-\alpha'\frac{3}{2q}} + ct^{\frac{1}{2}-\frac{S(b)}{2}-\alpha''\frac{3}{4q}}$$

Finally, we will finish the proof by showing that q can be chosen such that the powers in the last estimate can be estimated by $-\alpha$.

By choosing q close to $\frac{3}{2}$ and hence α' close to $\frac{1}{2}$, the second addend decays like $t^{-\frac{3}{4}+\varepsilon}$ for some small parameter $0 < \varepsilon < \frac{1}{4}$.

To consider the last power, we will prove that the power converges to a number strictly less than $-\alpha$ if we take the limit $q \searrow \frac{3}{2}$. Note that the first inequality in the next lines is valid due to the estimate $S(b) - \frac{5}{4} > \alpha$ and $q < \frac{q_2}{2}$. We obtain

$$\frac{1}{2} - \frac{S(b)}{2} - \alpha''\frac{3}{4q} = \frac{1}{2} - \frac{S(b)}{2} + \frac{9}{4q}\left(\frac{1}{q} - \frac{1}{q_2} + \frac{1}{6}\right)$$
$$\leq \frac{1}{2} - \frac{\alpha}{2} - \frac{5}{8} - \frac{9}{4q}\left(\frac{1}{2q} - \frac{1}{6}\right)$$
$$= -\frac{1}{8} - \frac{\alpha}{2} - \frac{9}{8q^2} + \frac{3}{8q} \rightarrow -\frac{3}{8} - \frac{\alpha}{2}$$
$$\leq -\alpha,$$

where we considered the limit $q \rightarrow \frac{3}{2}$, and thus also the third addend in (4.17) decays like $t^{-\alpha}$. The Theorem is proved. $\qquad\square$

Note that the last proof also shows that if $F = 0$ or $b = 0$ the same decay result holds and in the definition of α the term corresponding to F or b can be neglected. If $F = b = 0$ we note that even if $\alpha_1, \alpha_2 \geq \frac{3}{4}$ the best decay estimate we can prove, is that for all $\varepsilon > 0$ there exists a $c > 0$ such that

$$\|v(t)\|_2 \leq ct^{-\frac{3}{4}+\varepsilon}.$$

This is due to the fact that the term T_1 corresponding to $J_k v \otimes v$ can be estimated by

$$\|T_1\| \leq ct^{\frac{1}{4} - \frac{3}{2q}}$$

with $q > \frac{3}{2}$.

CHAPTER 5

Decay of Turbulent Solutions to the Navier–Stokes Equations

In the last chapter we considered the system (4.5) and proved that there exists a suitable solution which decays to 0 as $t \to \infty$ exponentially, if the domain is bounded, or polynomially, if Ω is an exterior domain. The aim of this chapter is now to prove that a large class of solutions has the same decay properties as the suitable one. To be more precise, we will prove that every *turbulent solution* tends to 0 as $t \to \infty$. The results in this section are published in [18].

For the integrability of the extension of the non-homogeneous boundary data, let us assume throughout this chapter that

(5.1)
$$b \in L^4(0, T; L^4(\Omega)) \cap L^{s_0}(0, T; L^{q_0}(\Omega)),$$
$$\frac{2}{s_0} + \frac{3}{q_0} = 1, \ s_0 \in (2, \infty).$$

Definition 5.0.1. Let $\Omega \subset \mathbb{R}^3$ be a domain with compact $C^{1,1}$ boundary and $0 < T \leq \infty$. Let b fulfil (5.1), let $f = \operatorname{div} F$ with $F \in L^2(0, T; L^2(\Omega))$ and $u_0 \in L^2_\sigma(\Omega)$. Then a vector field v on $(0, T) \times \Omega$ is called a *turbulent Leray–Hopf type weak solution to* (4.5), if v is a weak solution to (4.5) in the sense of Definition 4.1.3, and v fulfils the *strong energy inequality*

(5.2)
$$\frac{1}{2}\|v(t)\|_2^2 + \int_s^t \|\nabla v\|_2^2 d\tau$$
$$\leq \frac{1}{2}\|v(s)\|_2^2 - \int_s^t \langle F - (v+b) \otimes b, \nabla v\rangle d\tau$$

for almost all $s \in (0,\infty)$ and all $t \in (s,\infty)$.

Note that every turbulent solution is by definition a weak solution and thus (5.2) is fulfilled for $s = 0$.

The strategy to prove the decay results now is as follows. Let v denote a turbulent solution to (4.5). Then we will prove that v coincides eventually with a *strong solution* and this strong solution is unique in the class of weak solutions. Finally, we will prove that eventually there exists a suitable weak solution, which decays, and a simple argument will prove that those three solutions coincide.

5.1 Existence of Serrin-type Strong Solutions

Let us start with the definition of *Serrin-type strong solutions* to (4.5).

Definition 5.1.1. Let $\Omega \subset \mathbb{R}^3$ be a domain with compact $C^{1,1}$ boundary and $0 < T \leq \infty$. Let b fulfil (5.1), let $f = \operatorname{div} F$ with $F \in L^2(0,T;L^2(\Omega))$ and $u_0 \in L_\sigma^2(\Omega)$. Then a vector field v on $(0,T) \times \Omega$ is called a *Serrin-type strong solution to* (4.5), if v is a weak solution to (4.5) in the sense of Definition 4.1.3, and
$$v \in L^{s_0}(0,T;L^{q_0}(\Omega))$$

with s_0, q_0 as in (5.1).

Lemma 5.1.2. *A Serrin-type strong solution fulfils the energy equality*

(5.3)
$$\frac{1}{2}\|v(t)\|_2^2 + \int_0^t \|\nabla v\|_2^2 d\tau$$
$$= \frac{1}{2}\|u_0\|_2 - \int_0^t \langle F - (v+b) \otimes b, \nabla v\rangle d\tau$$

for all $t \in (0,T)$.

Proof. The regularity and integrability assumptions on a Serrin-type strong solution imply that

$$\langle (v+b) \otimes (v+b), \nabla v \rangle_{\Omega,T}$$
$$\leq \left(\|v^2\|_{2,2} + \|b^2\|_{2,2} + \|b\|_{q_0,s_0} \|v\|_{q,s} \right) \|\nabla v\|_{2,2}$$

with some q, s such that $\frac{2}{s} + \frac{3}{q} = \frac{3}{2}$ and $s \in (2, \infty)$. We obtain further

$$\langle (v+b) \otimes (v+b), \nabla v \rangle_{\Omega,T}$$
$$\leq c \big(\|v\|_{q_0,s_0} \|v\|_{q,s} + \|b\|_{4,4}^2 + \|b\|_{q_0,s_0} (\|v\|_{2,\infty} + \|\nabla v\|_{2,2}) \big) \|\nabla v\|_{2,2}$$
$$\leq c \big((\|v\|_{q_0,s_0} + \|b\|_{q_0,s_0})(\|v\|_{2,\infty} + \|\nabla v\|_{2,2}) + \|b\|_{4,4}^2 \big) \|\nabla v\|_{2,2}.$$

Here, we used the well-known inequality

$$\|v\|_{q,s} \leq \|v\|_{6,2} + \|v\|_{2,\infty} \leq c \|\nabla v\|_{2,2} + \|v\|_{2,\infty}$$

for $\frac{3}{q} + \frac{2}{s} = \frac{3}{2}$ for Leray–Hopf type weak solutions to Navier–Stokes equations. Let $(\varphi_n)_n \subset C_0^\infty([0,T); C_{0,\sigma}^\infty(\Omega))$ denote a sequence of test-functions that approximates v w.r.t. the norm of

$$L^2(0,T; \dot{W}_0^{1,2}(\Omega)) \cap L^{s_0}(0,T; L^{q_0}(\Omega)) \cap L^\infty(0,T; L^2(\Omega)).$$

Then by definition we obtain that for all $n \in \mathbb{N}$ we can insert φ_n as a test-function in (4.6), and the calculation above implies that we can pass to the limit. Finally, due to the cancelation property we obtain that

$$\langle v \otimes v, \nabla v \rangle_{\Omega,T} = 0$$

and hence we proved that v fulfils the energy equality (5.3) $\qquad\square$

Before we can state an existence result for Serrin-type strong solutions to the perturbed Navier–Stokes equations (4.5) we need to introduce the space for initial values. Therefore, let us fix $s \in (2, \infty)$ and $q \in (3, \infty)$ such that $2/s + 3/q = 1$. Furthermore, let us for short introduce the notation

$$X := L^s(0,T; L^q(\Omega)).$$

Then we define the space of initial values as

$$\mathcal{B} := \mathcal{B}_T^{q,s} := \{x \in \mathrm{L}_\sigma^2(\Omega) \mid \tau \mapsto e^{-\tau A}x \in X\}$$

endowed with the norm

$$\|x\|_{\mathcal{B}_T^{q,s}} := \Big(\int_0^T \|e^{-\tau A}x\|_q^s \mathrm{d}s\Big)^{\frac{1}{s}}.$$

Furthermore, for the existence theorem of strong solutions, we need to assume that

$$(5.4) \qquad b \in \mathrm{L}^{s_1}(0,T;\mathrm{L}^{q_1}(\Omega)), \ \frac{1}{s_0}+\frac{1}{s_1}=\frac{1}{q_0}+\frac{1}{q_1}=\frac{1}{2}.$$

Theorem 5.1.3. *Let $\Omega \subset \mathbb{R}^3$ be a domain with compact $C^{1,1}$ boundary and $0 < T \leq \infty$. Let b fulfil (5.1) as well as (5.4), let $f = \operatorname{div} F$ with $F \in \mathrm{L}^2(0,T;\mathrm{L}^2(\Omega))$, and $u_0 \in \mathrm{L}_\sigma^2(\Omega)$. Furthermore, if Ω is an exterior domain, assume that $q_0 > 6$ in (5.1). Assume furthermore that*

$$\|u_0\|_{\mathcal{B}_T^{q_0,s_0}} < \infty$$

and assume

$$F \in \mathrm{L}^{\frac{s_0}{2}}(0,T;\mathrm{L}^{\frac{q_0}{2}}(\Omega)).$$

Then there exists a $\varepsilon > 0$ with the following property: For any $0 < T' \leq T$ such that

$$\|F\|_{\mathrm{L}^{\frac{s_0}{2}}(0,T';\mathrm{L}^{\frac{q_0}{2}}(\Omega))} + \|b\|_{\mathrm{L}^{s_0}(0,T';\mathrm{L}^{q_0}(\Omega))} + \|u_0\|_{\mathcal{B}_{T'}^{q_0,s_0}} < \varepsilon,$$

there exists a unique Serrin-type strong solution to (4.5) with

$$u \in \mathrm{L}^{s_0}(0,T';\mathrm{L}^{q_0}(\Omega)).$$

Before we start with the proof of Theorem 5.1.3, let us prove the following preliminary result.

Proposition 5.1.4. *Let* $\Omega \subset \mathbb{R}^3$ *be a domain with compact* $C^{1,1}$ *boundary and* $0 < T \leq \infty$. *Let* b *fulfil* (5.1) *and* (5.4), *let* $f = \operatorname{div} F$ *for some* $F \in \mathrm{L}^2(0,T;\mathrm{L}^2(\Omega))$, *and* $u_0 \in \mathrm{L}^2_\sigma(\Omega)$. *Assume furthermore that*

$$\|u_0\|_{\mathcal{B}_T^{q_0,s_0}} < \infty$$

and assume

$$F \in \mathrm{L}^{\frac{s_0}{2}}(0,T;\mathrm{L}^{\frac{q_0}{2}}(\Omega)).$$

Then there exists a $\varepsilon > 0$ *with the following property: For any* $0 < T' \leq T$ *such that*

$$\|F\|_{\mathrm{L}^{\frac{s_0}{2}}(0,T';\mathrm{L}^{\frac{q_0}{2}}(\Omega))} + \|b\|_{\mathrm{L}^{s_0}(0,T';\mathrm{L}^{q_0}(\Omega))} + \|u_0\|_{\mathcal{B}_{T'}^{q_0,s_0}} < \varepsilon,$$

the operator

$$\mathcal{F}\colon \mathrm{L}^{s_0}(0,T';\mathrm{L}^{q_0}(\Omega)) \to \mathrm{L}^{s_0}(0,T';\mathrm{L}^{q_0}(\Omega)),$$

$$w \mapsto -\int_0^t A^{\frac{1}{2}} e^{-(t-\tau)A}(A^{-\frac{1}{2}}P\operatorname{div})((w+v_0+b)(w+v_0+b))\mathrm{d}\tau,$$

where

$$v_0(t) := e^{-tA}u_0 + \int_0^t A^{\frac{1}{2}} e^{-(t-\tau)A}(A^{-\frac{1}{2}}P\operatorname{div})F(\tau)\mathrm{d}\tau,$$

has a fixed point.

Proof. Let $X_{T'} := \mathrm{L}^{s_0}(0,T';\mathrm{L}^{q_0}(\Omega))$ for any $0 < T' \leq T$ and let us define

$$\begin{aligned}
v_0(t) &= e^{-tA}u_0 + \int_0^t A^{\frac{1}{2}} e^{-(t-\tau)A}(A^{-\frac{1}{2}}P\operatorname{div})F(\tau)\mathrm{d}\tau \\
&=: V_0(t) + V_1(t).
\end{aligned}$$

The assumption $u_0 \in \mathcal{B}_T^{q_0,s_0}$ implies that $V_0 \in X_T$ and that

$$\|V_0\|_{X_{T'}} \leq \|u_0\|_{\mathcal{B}_{T'}^{q_0,s_0}}$$

for all $0 < T' \leq T$. To consider V_1 we estimate

$$\begin{aligned}
\|V_1(t)\|_{q_0} &\leq \int_0^t \|A^{\frac{1}{2}} e^{-(t-\tau)A}(A^{-\frac{1}{2}}P\operatorname{div})F(\tau)\|_{q_0}\mathrm{d}\tau \\
&\leq c \int_0^t (t-\tau)^{-\frac{1}{2}-\frac{3}{2q_0}} \|(A^{-\frac{1}{2}}P\operatorname{div})F(\tau)\|_{\frac{q_0}{2}}\mathrm{d}\tau \\
&\leq c \int_0^t (t-\tau)^{-\frac{1}{2}-\frac{3}{2q_0}} \|F(\tau)\|_{\frac{q_0}{2}}\mathrm{d}\tau,
\end{aligned}$$

where we used (1.24) and the boundedness of $A^{-\frac{1}{2}}P\mathrm{div}$ in $\mathrm{L}^{\frac{q_0}{2}}(\Omega)$. The Hardy–Littlewood–Sobolev inequality implies

$$\|V_1\|_{q_0,s_0,T'} \leq c\|F\|_{\frac{q_0}{2},\frac{s_0}{2},T'}.$$

For an arbitrary $w \in X$ we obtain with the same estimates as above that

$$\|\mathcal{F}w\|_{q_0,s_0,T'} \leq c\|(w+v_0+b)(w+v_0+b)\|_{q_0/2,s_0/2,T'}$$
$$\leq c\|w+v_0+b\|_{q_0,s_0,T'}^2.$$

Thus, $\mathcal{F}$ maps $X_{T'}$ into $X_{T'}$ and

$$\|\mathcal{F}(w)\|_{q_0,s_0,T'} \leq c(\|w\|_{q_0,s_0,T'} + \beta(T'))^2$$

with $\beta(T') = \|v_0\|_{q_0,s_0,T'} + \|b\|_{q_0,s_0,T'}$. By analogy, for given $w_1, w_2 \in X$, we obtain that

$$\|\mathcal{F}(w_1) - \mathcal{F}(w_2)\|_{q_0,s_0,T'}$$
$$\leq c\|w_1 - w_2\|_{q_0,s_0,T'}\big(\|w_1+v_0+b\|_{q_0,s_0,T'} + \|w_2+v_0+b\|_{q_0,s_0,T'}\big)$$
$$\leq c\|w_1 - w_2\|_{q_0,s_0,T'}\big(\|w_1\|_{q_0,s_0,T'} + \|w_2\|_{q_0,s_0,T'} + 2\beta(T')\big).$$

Choosing $T' \in (0,T]$ sufficiently small such that $4c\beta(T') < 1$, we can apply Proposition 3.1.1 and obtain the existence of a fixed point. $\qquad\Box$

Let us now turn to the proof of Theorem 5.1.3.

Proof of Theorem 5.1.3. Let v_0 as in Proposition 5.1.4 and w denote a fixed point of $\mathcal{F}$, where $\mathcal{F}$ is defined as in Proposition 5.1.4. Note that $v := v_0 + w$ fulfils

(5.5)
$$v(t) = e^{-tA}u_0 + \int_0^t A^{\frac{1}{2}}e^{-(t-\tau)A}(A^{-\frac{1}{2}}P\mathrm{div})(F - (v+b)(v+b))(\tau)\,\mathrm{d}\tau,$$

and hence v is a weak solution to (4.5) if v is in the Leray–Hopf class.

The assumptions on F, b imply that $\tilde{F} := F - vb - bv - bb \in \mathrm{L}^2(0,T';\mathrm{L}^2(\Omega))$. Thus, we obtain that

$$\tilde{v}(t) := e^{-tA}u_0 + \int_0^t A^{\frac{1}{2}}e^{-(t-\tau)A}A^{-\frac{1}{2}}P\mathrm{div}\,\tilde{F}(\tau)\,\mathrm{d}\tau$$
$$= v(t) - \int_0^t A^{\frac{1}{2}}e^{-(t-\tau)A}A^{-\frac{1}{2}}P\mathrm{div}\,vv\,\mathrm{d}\tau$$

is a weak solution to the nonstationary Stokes equations with initial value u_0 and right-hand side div $\tilde{F}$, see [50, Ch. V, Theorem 2.4.1]; in particular, $\nabla\tilde{v} \in L^2(0, T'; L^2(\Omega))$. Thus $\tilde{v}$ is in the Leray–Hopf class with

$$\|\tilde{v}\|_{q_1,s_1,T'} \leq c(\|\tilde{v}\|_{2,\infty,T'} + \|\nabla\tilde{v}\|_{2,2,T'}),$$

where s_1, q_1 are defined as in (5.4) and consequently satisfy $\frac{2}{s_1} + \frac{3}{q_1} = \frac{3}{2}$.

Let us prove next that $\nabla v \in L^2(0, T'; L^2(\Omega))$. Therefore, let

$$J_n := (1 + \frac{1}{n}A^{\frac{1}{2}})^{-1}$$

denote the Yosida operator and let $v_n = J_n v$. We have

$$J_n P\text{div}\,(vv) = J_n P\text{div}\,(v(J_n^{-1}v_n))$$
$$= J_n P(v \cdot \nabla v_n) + \left(\frac{1}{n}J_n A^{\frac{1}{2}}\right)(A^{-\frac{1}{2}}P\text{div}\,)(v(A^{\frac{1}{2}}v_n)).$$

Using the boundedness of $A^{-\frac{1}{2}}P\text{div}\,$, of J_n, and of $\frac{1}{n}A^{\frac{1}{2}}J_n$ uniformly in $n \in \mathbb{N}$, as well as (1.20) we obtain with $\frac{1}{\gamma} = \frac{1}{2} + \frac{1}{q_0}$ that

$$\|J_n P\text{div}\,(vv)(t)\|_\gamma \leq c\|v(t)\|_{q_0}\|A^{\frac{1}{2}}v_n(t)\|_2.$$

If Ω is an exterior domain, the boundedness of $A^{-\frac{1}{2}}P\text{div}\,$ in $L^\gamma(\Omega)$ requires that $\gamma > \frac{3}{2}$ or equivalently $q_0 > 6$. Due to (1.23) with $\alpha = \frac{3}{2\gamma} - \frac{3}{4}$ and (1.24) we get for $A^{\frac{1}{2}}v_n$, see (5.5), the estimate

$$\|A^{\frac{1}{2}}v_n(t)\|_2 \leq \|A^{\frac{1}{2}}J_n\tilde{v}(t)\|_2 + c\int_0^t \|A^{\frac{1}{2}+\alpha}e^{-(t-\tau)A}J_n P\text{div}\,(vv)(\tau)\|_\gamma \, d\tau$$
$$\leq \|A^{\frac{1}{2}}J_n\tilde{v}(t)\|_2 + c\int_0^t (t-\tau)^{-1+\frac{1}{s_0}}(\|v(\tau)\|_{q_0}\|A^{\frac{1}{2}}v_n(\tau)\|_2) \, d\tau.$$

Thus, the Hardy–Littlewood–Sobolev inequality yields the estimate

$$\|A^{\frac{1}{2}}v_n\|_{2,2,T'} \leq \|A^{\frac{1}{2}}J_n\tilde{v}\|_{2,2,T'} + c\|v\|_{q_0,s_0,T'}\|A^{\frac{1}{2}}v_n\|_{2,2,T'}.$$

By choosing ε sufficiently small enough we can assume that $c\|v\|_{q_0,s_0,T'} \leq \frac{1}{2}$ and hence by an absorption argument we obtain that

$$\|A^{\frac{1}{2}}v_n\|_{2,2,T'} \leq 2\|A^{\frac{1}{2}}J_n\tilde{v}\|_{2,2,T'} \leq c\|\nabla\tilde{v}\|_{2,2,T'}.$$

Since the last estimate is independent of n, a reflexivity argument implies $A^{\frac{1}{2}}v \in L^2(0, T'; L^2(\Omega))$.

Next, let us prove that $vv \in L^2(0, T'; L^2(\Omega))$. Therefore, let

$$\bar{v}(t) := \int_0^t e^{-(t-\tau)A} P \mathrm{div}\,(vv)\,\mathrm{d}\tau$$

and choose s_2, q_2 such that

$$\frac{1}{q_2} = \frac{1}{2} + \frac{1}{q_0}, \quad \frac{1}{s_2} = \frac{1}{2} + \frac{1}{s_0}.$$

Then we may estimate, using also q_1, s_1 as in (5.4),

$$\|\bar{v}(t)\|_{q_1} \leq c \int_0^t \|A^{\frac{3}{q_0}} e^{-(t-\tau)A} P(v \cdot \nabla v)(\tau)\|_{q_2}\,\mathrm{d}\tau$$
$$\leq c \int_0^t (t-\tau)^{-\frac{3}{q_0}} \|v\|_{q_0} \|\nabla v\|_2\,\mathrm{d}\tau.$$

Therefore, the Hardy–Littlewood–Sobolev inequality implies that

$$\|\bar{v}\|_{q_1,s_1,T'} \leq c\|v\|_{q_0,s_0,T'} \|\nabla v\|_{2,2,T'},$$

and we conclude that $v = \tilde{v} + \bar{v} \in L^{s_1}(0, T'; L^{q_1}(\Omega))$. Thus,

$$\|vv\|_{2,2,T'} \leq \|v\|_{q_1,s_1,T'} \|v\|_{q_0,s_0,T'} < \infty.$$

As for $\tilde{v}$ above, we obtain that v is a weak solution to the Stokes equations with right-hand side $F - vv - bv - vb - bb \in L^2(0, T'; L^2(\Omega))$ and thus v fulfils the energy equality (5.3). Especially, $v \in L^\infty(0, T'; L^2(\Omega))$ and hence v is in the Leray–Hopf class. In summary, we have proved Theorem 5.1.3 except for the uniqueness of the strong solution. The uniqueness will be shown in the next section. $\qquad\square$

5.2 On Serrin's Uniqueness Condition

In this section we are going to prove Theorem 5.2.1, *i.e.*, a version of Serrin's Uniqueness Theorem [47] for the equation (4.5).

Theorem 5.2.1. *Let* $\Omega \subset \mathbb{R}^3$ *be a domain with compact* $C^{1,1}$ *boundary. Let* b *fulfil* (5.1), *let* $f = \operatorname{div} F$, $F \in \mathrm{L}^2(0,T;\mathrm{L}^2(\Omega))$, *and* $u_0 \in \mathrm{L}^2_\sigma(\Omega)$. *Finally, let* v *be a Leray–Hopf type weak solution to* (4.5) *and* w *be a Serrin-type strong solution to* (4.5). *Then* $v = w$.

To prove this theorem we will follow the ideas of [50, Ch. V, 1.5]. Throughout this section let u denote a Leray–Hopf type weak solution to (4.5) and let $w \in \mathrm{L}^{s_0}(0,T;\mathrm{L}^{q_0}(\Omega))$, $0 < T \leq \infty$, denote a strong Serrin-type strong solution to (4.5). Furthermore, let $\rho\colon \mathbb{R} \to \mathbb{R}$ be a mollifier, *i.e.*, $0 \leq \rho \in C_0^\infty(\mathbb{R})$ is even and $\int \rho(t)\,\mathrm{d}t = 1$. Then we define the convolution operator in time by

$$v_\varepsilon(t) := \int_0^T \frac{1}{\varepsilon}\rho\Big(\frac{t-\tau}{\varepsilon}\Big)v(\tau)\,\mathrm{d}\tau, \quad \varepsilon > 0,$$

for a given $v \in \mathrm{L}^s(0,T;\mathrm{L}^q(\Omega))$, $1 \leq s, q \leq \infty$. All convergence results needed in the proof are stated in the following Lemma 5.2.2.

Lemma 5.2.2. *Choose* $s_1, q_1 \in (1,\infty)$ *such that*

(5.6)
$$\frac{1}{2} = \frac{1}{s_1} + \frac{1}{s_0} = \frac{1}{q_1} + \frac{1}{q_0}$$

and let q_0', s_0' *denote the conjugate exponents of* q_0, s_0. *Then*

$$\|\nabla u_\varepsilon - \nabla u\|_{2,2} + \|\nabla w_\varepsilon - \nabla w\|_{2,2} + \|w_\varepsilon - w\|_{q_0,s_0} + \|F_\varepsilon - F\|_{2,2}$$
$$+ \|(bu)_\varepsilon - bu\|_{2,2} + \|(bw)_\varepsilon - bw\|_{2,2}$$
$$+ \|(ww)_\varepsilon - ww\|_{2,2} + \|\operatorname{div}(uu)_\varepsilon - \operatorname{div} uu\|_{q_0',s_0'} \to 0 \ as \ \varepsilon \to 0.$$

Furthermore, for all $t \in (0,T)$, *it holds that*

$$w_\varepsilon(t) \to w(t) \ in \ \mathrm{L}^2(\Omega), \quad u_\varepsilon(t) \rightharpoonup u(t) \ in \ \mathrm{L}^2(\Omega).$$

Proof. Most of these results on convergence can be proved easily with a standard mollifier argument. Let us just mention that $\frac{2}{s_1} + \frac{3}{q_1} = \frac{3}{2}$ and $\frac{1}{2} + \frac{1}{s_1} = \frac{1}{s_0'}$, $\frac{1}{2} + \frac{1}{q_1} = \frac{1}{q_0'}$. Hence

$$\|\operatorname{div}(uu)\|_{q_0',s_0'} \leq \|u\|_{q_1,s_1}\|\nabla u\|_{2,2} \leq c(\|u\|_{2,\infty} + \|\nabla u\|_{2,2})\|\nabla u\|_{2,2} < \infty$$

and

$$\|ww\|_{2,2} \le \|w\|_{q_0,s_0} \|w\|_{q_1,s_1} < \infty$$
$$\|bu\|_{2,2} \le \|b\|_{q_0,s_0} \|u\|_{q_1,s_1} < \infty.$$

The statement of the pointwise convergence follows from the (weak) L^2-continuity in time of u or w, respectively. $\qquad\qquad\qquad\qquad$ □

Let us start to prove Theorem 5.2.1.

Proof of Theorem 5.2.1. Let $0 < t_0 < t_1 < T$ and let $\varphi \in C_0^\infty((t_0, t_1))$. Using equation (4.6) for u with the test function $(\varphi w_\varepsilon)_\varepsilon$ we get an integral identity of the form

$$-\int_0^T \langle u(\tau), ((\varphi w_\varepsilon)_\varepsilon)_t(\tau)\rangle \, d\tau = \int_0^T \langle \dots, (\varphi \nabla w_\varepsilon)_\varepsilon(\tau)\rangle \, d\tau,$$

where the left-hand term may be rewritten as $\int_{t_0}^{t_1} \langle \partial_t u_\varepsilon, w_\varepsilon\rangle \varphi \, d\tau$ due to supp $\varphi \subset (t_0, t_1)$ and ρ is even. Since this identity holds for arbitrary $\varphi \in C_0^\infty((t_0, t_1))$, we conclude that the equality

(5.7)
$$\langle \partial_t u_\varepsilon(t), w_\varepsilon(t)\rangle + \langle \nabla u_\varepsilon(t), \nabla w_\varepsilon(t)\rangle - \langle ((u+b)(u+b))_\varepsilon(t), \nabla w_\varepsilon(t)\rangle$$
$$= -\langle F_\varepsilon(t), \nabla w_\varepsilon(t)\rangle$$

holds for all $t \in (t_0, t_1)$. Adding equation (5.7) and the corresponding equation (4.6) for w tested with $(\varphi u_\varepsilon)_\varepsilon$, integrating from t_0 to t_1, and using the simple identity

$$\int_{t_0}^{t_1} \langle \partial_t u_\varepsilon(\tau), w_\varepsilon(\tau)\rangle + \langle u_\varepsilon(\tau), \partial_t w_\varepsilon(\tau)\rangle \, d\tau = \langle u_\varepsilon(t_1), w_\varepsilon(t_1)\rangle - \langle u_\varepsilon(t_0), w_\varepsilon(t_0)\rangle,$$

we obtain

$$\langle u_\varepsilon(t_1), w_\varepsilon(t_1)\rangle - \langle u_\varepsilon(t_0), w_\varepsilon(t_0)\rangle + 2\int_{t_0}^{t_1} \langle \nabla u_\varepsilon(\tau), \nabla w_\varepsilon(\tau)\rangle \, d\tau$$
$$-\int_{t_0}^{t_1} \langle ((u+b)(u+b))_\varepsilon(\tau), \nabla w_\varepsilon(\tau)\rangle + \langle ((w+b)(w+b))_\varepsilon(\tau), \nabla u_\varepsilon(\tau)\rangle \, d\tau$$
$$= -\int_{t_0}^{t_1} \langle F_\varepsilon(\tau), \nabla u_\varepsilon(\tau) + \nabla w_\varepsilon(\tau)\rangle \, d\tau.$$

Taking the limit $\varepsilon \to 0$ as well as $t_0 \to 0$ we get

(5.8)
$$
\begin{aligned}
&\langle u(t_1), w(t_1)\rangle + 2\int_0^{t_1} \langle \nabla u(\tau), \nabla w\rangle \, \mathrm{d}\tau \\
&- \int_0^{t_1} \langle (u+b)(u+b), \nabla w(\tau)\rangle + \langle (w+b)(w+b), \nabla u\rangle \, \mathrm{d}\tau \\
&= -\int_0^{t_1} \langle F, \nabla u + \nabla w\rangle \, \mathrm{d}\tau + \|u_0\|_2^2.
\end{aligned}
$$

Note that in (5.8) the term $\langle bu, \nabla w\rangle + \langle bw, \nabla u\rangle = 0$. Adding the energy inequality (4.7) for u, the energy equality (5.3) for w and subtracting (5.8), we see that

$$
\begin{aligned}
0 \geq\; & \frac{1}{2}\|u(t_1)\|_2^2 + \frac{1}{2}\|w(t_1)\|_2^2 - \langle u(t_1), w(t_1)\rangle \\
& + \int_0^{t_1} \|\nabla u\|_2^2 + \|\nabla w\|_2^2 - 2\langle \nabla u, \nabla w\rangle \, \mathrm{d}\tau \\
& - \int_0^{t_1} \langle (u+b)b, \nabla u\rangle + \langle (w+b)b, \nabla w\rangle \, \mathrm{d}\tau \\
& + \int_0^{t_1} \langle (u+b)(u+b), \nabla w\rangle + \langle (w+b)(w+b), \nabla u\rangle \, \mathrm{d}\tau \\
=\; & \frac{1}{2}\|u(t_1) - w(t_1)\|_2^2 + \int_0^{t_1} \|\nabla(u-w)\|_2^2 \, \mathrm{d}\tau \\
& - \int_0^{t_1} \langle (u-w)b, \nabla(u-w)\rangle - \langle uu, \nabla w\rangle - \langle ww, \nabla u\rangle \, \mathrm{d}\tau \\
=\; & \frac{1}{2}\|W(t_1)\|_2^2 + \int_0^{t_1} \|\nabla W\|_2^2 \, \mathrm{d}\tau \\
& - \int_0^{t_1} \langle Wb, \nabla W\rangle - \langle uu, \nabla w\rangle - \langle ww, \nabla u\rangle \, \mathrm{d}\tau
\end{aligned}
$$

with $W := u - w$. Note that

$$
\langle uu, \nabla w\rangle + \langle ww, \nabla u\rangle = -\langle W \cdot \nabla W, w\rangle
$$

and hence

$$
\frac{1}{2}\|W(t)\|_2^2 + \int_0^{t_1} \|\nabla W(\tau)\|_2^2 \, \mathrm{d}\tau \leq \int_0^{t_1} \langle Wb, \nabla W\rangle - \langle W \cdot \nabla W, w\rangle \, \mathrm{d}\tau.
$$

Let us define the energy norm $\|\cdot\|_{t_1}$ by

$$
\|W\|_{t_1}^2 := \frac{1}{2} \sup_{0<t<t_1} \|W(t)\|_2^2 + \int_0^{t_1} \|\nabla W(\tau)\|_2^2 \, \mathrm{d}\tau.
$$

Using (5.6) and the estimate $\|W\|_{q_1,s_1,t_1} \le c\|W\|_{t_1}$ we get that

$$
\begin{aligned}
\|W\|_{t_1}^2 &\le \int_0^{t_1} \langle Wb, \nabla W \rangle - \langle w, W \cdot \nabla W \rangle \, d\tau \\
&\le (\|b\|_{q_0,s_0,t_1} + \|w\|_{q_0,s_0,t_1}) \|\nabla W\|_{2,2,t_1} \|W\|_{q_1,s_1,t_1} \\
&\le c(\|b\|_{q_0,s_0,t_1} + \|w\|_{q_0,s_0,t_1}) \|W\|_{t_1}^2 .
\end{aligned}
$$

Choosing t_1 such that $c(\|b\|_{q_0,s_0,t_1} + \|w\|_{q_0,s_0,t_1}) < 1$ we see that $W_{|(0,t_1)} = 0$. Considering the shifted function $w(\cdot - t_1)$, $u(\cdot - t_1)$ and using an induction argument one proves easily that $W = 0$ and hence $u = w$. $\qquad\square$

5.3 Decay of Turbulent Solutions

In this final section of this chapter, we will prove the decay results for arbitrary turbulent solutions to (4.5). As presented in the beginning of this chapter, the idea is to prove that the turbulent solution u coincides with a strong solution in a time interval $[T_0, \infty)$ for some suitable $T_0 \ge 0$. Let us start to prove that there exists a positive time T_0 such that $u(T_0)$ is a suitable initial value for Serrin-type strong solution.

Lemma 5.3.1. *Let $\Omega \subset \mathbb{R}^3$ be a domain with compact $C^{1,1}$-boundary and let $x \in L_\sigma^2(\Omega) \cap L^6(\Omega)$. Let furthermore*

$$
\frac{2}{s} + \frac{3}{q} = 1, \ s \in (2,\infty), \ q \in (3,\infty).
$$

Then there exists an $\alpha > 0$ and a $c > 0$ such that

(5.9) $$\|x\|_{\mathcal{B}_\infty^{q,s}} \le c\|x\|_2^{1-\alpha} \|x\|_6^{\alpha}.$$

Proof. Let us first assume that $q \ge 6$. Then we have for arbitrary $R > 0$ that

$$
\begin{aligned}
\int_0^\infty \|e^{-tA}x\|_q^s dt &= \int_0^R \|e^{-tA}x\|_q^s dt + \int_R^\infty \|e^{-tA}x\|_q^s dt \\
&\le c \int_0^R t^{-\frac{3}{2}\left(\frac{1}{6}-\frac{1}{q}\right)s} \|x\|_6^s dt + c \int_R^\infty t^{-\frac{3}{2}\left(\frac{1}{2}-\frac{1}{q}\right)s} \|x\|_2^s dt \\
&\le c\left(R^{\frac{s}{4}} \|x\|_6^s + R^{-\frac{s}{4}} \|x\|_2^s\right)
\end{aligned}
$$

using (1.24). Choosing $R = (\|x\|_2 \|x\|_6^{-1})^2$ yields the result.

In the case $q \in (3,6)$ let us choose $\beta > 0$ such that $\frac{1}{q} = \frac{\beta}{6} + \frac{1-\beta}{2}$. Then we have

$$\int_0^\infty \|e^{-tA}x\|_q^s \, dt \le \int_0^R \|e^{-tA}x\|_2^{(1-\beta)s} \|e^{-tA}x\|_6^{\beta s} \, dt + c\int_R^\infty t^{-\frac{s}{4}-1}\|x\|_2^s \, dt$$
$$\le c\|x\|_2^{(1-\beta)s}\Big(R\|x\|_6^{\beta s} + R^{-\frac{s}{4}}\|x\|_2^{\beta s}\Big).$$

Now choosing R such that both terms in the bracket can be compared, hence
$$R^{1+\frac{s}{4}} = (\|x\|_2 \|x\|_6^{-1})^{\beta s},$$

we obtain
$$\int_0^\infty \|e^{-tA}x\|_q^s \, dt \le c\|x\|_2^{(1-\beta)s}(\|x\|_2 \|x\|_6^{\frac{s}{4}})^{\frac{\beta s}{1+s/4}}$$

and thus the statement is proved. $\qquad\square$

Let us now state and prove the main theorems of this chapter.

Theorem 5.3.2. *Let $\Omega \subset \mathbb{R}^3$ be a bounded domain with $C^{1,1}$-boundary. Furthermore, let b fulfil (5.1) and (5.4) with $T = \infty$, let $f = \operatorname{div} F$, $F \in L^2(0,\infty; L^2(\Omega))$. Furthermore, assume that there exists a $T > 0$ such that*

$$F \in L^{\frac{s_0}{2}}(T, \infty; L^{\frac{q_0}{2}}(\Omega)).$$

Then for every turbulent solution v to (4.5) we have

$$\lim_{t\to\infty} \|v(t)\|_2 = 0.$$

Let $\alpha \in (0,1)$, $\beta > 0$, and assume in addition that

(5.10) $\quad \|F\|_{L^2(\alpha t, t; L^2(\Omega))}^2 + \|b\|_{L^4(\alpha t, t; L^4(\Omega))}^4 + \|b\|_{L^{s_0}(\alpha t, t; L^{q_0}(\Omega))}^{s_0} = O(e^{-\beta t})$

as $t \to \infty$. Then for every turbulent solution v to (4.5) we have

$$\|v(t)\|_2 \in O(\exp(-t\gamma)) \ as \ t \to \infty.$$

here, $\gamma := \frac{1}{2}\min\{(1-\alpha)\rho, \beta\}$ and $\sqrt[-2]{\rho}$ denotes the largest constant, such that Poincaré's inequality holds, i.e., $\rho\|u\|_2^2 \le \|\nabla u\|_2^2$ for all $u \in W_0^{1,2}(\Omega)$.

Proof. Note that v is in the Leray–Hopf class, *i.e.*, $v \in L^\infty(0, \infty; L^2(\Omega))$ and $\nabla v \in L^2(0, \infty; L^2(\Omega))$ and due to the embedding $\|x\|_6 \leq c\|\nabla x\|_2$ we obtain

$$v \in L^\infty(0, \infty; L^2(\Omega)) \cap L^2(0, \infty; L^6(\Omega)).$$

Due to Lemma 5.3.1 and since v fulfils the strong energy inequality (5.2) we obtain that for every $t_0 \geq 0$ and every $\delta > 0$ there exists a $t_1 \geq t_0$ such that

$$\|v(t_1)\|_{B^{q_0, s_0}_\infty} < \delta.$$

Hence due to Theorem 5.1.3 we obtain that there exists a $T_0 \geq 0$ such that

$$\|F\|_{L^{\frac{s_0}{2}}(T_0, \infty; L^{\frac{q_0}{2}}(\Omega))} + \|b\|_{L^{s_0}(T_0, \infty; L^{q_0}(\Omega))} + \|u(T_0)\|_{B^{q_0, s_0}_\infty} < \varepsilon,$$

where ε denotes the constant in Theorem 5.1.3. This implies that there exists a unique Serrin-type strong solution w in the time interval $[T_0, \infty)$ and Theorem 5.2.1 implies $v_{|[T_0, \infty)} = w$. Furthermore, due to Theorem 4.2.1 there exists a solution $\tilde{v}$ in the time interval $[T_0, \infty)$ which decays in a suitable manner. And, also due to Theorem 5.2.1, we have $v = w = \tilde{v}$. Hence v decays and the Theorem is proved. $\qquad\square$

    With the same strategy we can also prove a decay result for arbitrary turbulent solution in the exterior domain case. The idea is always to show that eventually the turbulent solutions coincides with a strong solution and a solution which decays in a suitable way. Since the proof is word by word the same than the proof of Theorem 5.3.2, we will omit the proof.

Theorem 5.3.3. *Let $\Omega \subset \mathbb{R}^3$ be an exterior domain with $C^{1,1}$-boundary. Furthermore, let b fulfil (5.1) with $q_0 > 6$ and (5.4) with $T = \infty$, let $f = \operatorname{div} F$, $F \in L^2(0, \infty; L^2(\Omega))$. Furthermore, assume that there exists a $T > 0$ such that*

$$F \in L^{\frac{s_0}{2}}(T, \infty; L^{\frac{q_0}{2}}(\Omega))$$

for some $T \geq 0$.

$$\nabla b \in \mathrm{L}^{s_1}(0, \infty; \mathrm{L}^{q_1}(\Omega)), \quad \frac{2}{s_1} + \frac{3}{q_1} = 2, \; q_1 > 3,$$

$$b \in \mathrm{L}^{s_2}(0, \infty; \mathrm{L}^{q_2}(\Omega)), \quad \frac{2}{s_2} + \frac{3}{q_2} =: S(b), \; 3 < q_2 \leq 4, \; s_2 \geq 2,$$

$$b \cdot \nabla b \in \mathrm{L}^1(0, \infty; \mathrm{L}^2(\Omega)),$$

$$\nabla F \in \mathrm{L}^1(0, \infty; \mathrm{L}^2(\Omega)),$$

$$F \in \mathrm{L}^{s_3}(0, \infty; \mathrm{L}^{q_3}(\Omega)), \quad \frac{2}{s_3} + \frac{3}{q_3} =: S(F), \; \frac{3}{2} < q_3 \leq 2, \; s_3 \geq 1,$$

and that

$$\|\nabla F\|_{\mathrm{L}^1(t/2, t, \mathrm{L}^2(\Omega))} + \|\nabla b\|_{\mathrm{L}^{s_1}(t/2, t, \mathrm{L}^{q_1}(\Omega))} + \|b \cdot \nabla b\|_{\mathrm{L}^1(t/2, t, \mathrm{L}^2(\Omega))} \in O(t^{-\alpha_2})$$

as $t \to \infty$ for some $\alpha_2 > 0$. Moreover, suppose that

$$\|e^{-tA} u_0\|_2 \in O(t^{-\alpha_1})$$

for some $\alpha_1 \geq 0$. Then for every turbulent solution v to (4.5) we obtain that

$$\|v(t)\|_2 \in O(t^{-\alpha}) \; as \; t \to \infty,$$

where

$$\alpha := \min \Big\{ S(b) - \frac{5}{4}, \frac{S(F)}{2} - \frac{5}{4}, \alpha_1, \alpha_2 \Big\}.$$

Bibliography

[1] H. Amann. *Linear and quasilinear parabolic problems. Vol. I, Abstract linear theory*, volume 89 of *Monographs in Mathematics*. Birkhäuser Boston, Inc., Boston, MA, 1995.

[2] H. Amann. Navier-Stokes equations with nonhomogeneous Dirichlet data. *J. Nonlinear Math. Phys.*, 10(suppl. 1):1–11, 2003.

[3] J. Bergh and J. Löfström. *Interpolation spaces. An introduction.* Springer-Verlag, Berlin-New York, 1976. Grundlehren der Mathematischen Wissenschaften, No. 223.

[4] W. Borchers and T. Miyakawa. Algebraic L^2 decay for Navier-Stokes flows in exterior domains. *Acta Math.*, 165(3-4):189–227, 1990.

[5] W. Borchers and T. Miyakawa. Algebraic L^2 decay for Navier-Stokes flows in exterior domains. II. *Hiroshima Math. J.*, 21(3):621–640, 1991.

[6] W. Borchers and H. Sohr. On the semigroup of the Stokes operator for exterior domains in L^q-spaces. *Math. Z.*, 196(3):415–425, 1987.

[7] J. Bourgain. Some remarks on Banach spaces in which martingale difference sequences are unconditional. *Ark. Mat.*, 21(2):163–168, 1983.

[8] D. L. Burkholder. A geometrical characterization of Banach spaces in which martingale difference sequences are unconditional. *Ann. Probab.*, 9(6):997–1011, 1981.

[9] P. L. Butzer and H. Berens. *Semi-groups of operators and approximation.* Die Grundlehren der mathematischen Wissenschaften, Band 145. Springer-Verlag New York Inc., New York, 1967.

[10] R. Denk, M. Hieber, and J. Prüss. $\mathscr{R}$-boundedness, Fourier multipliers and problems of elliptic and parabolic type. *Mem. Amer. Math. Soc.*, 166(788):viii+114, 2003.

[11] G. Dore and A. Venni. On the closedness of the sum of two closed operators. *Math. Z.*, 196(2):189–201, 1987.

[12] K.-J. Engel and R. Nagel. *One-parameter semigroups for linear evolution equations*, volume 194 of *Graduate Texts in Mathematics.* Springer-Verlag, New York, 2000.

[13] J. Escher, J. Prüss, and G. Simonett. Analytic solutions for a Stefan problem with Gibbs-Thomson correction. *J. Reine Angew. Math.*, 563:1–52, 2003.

[14] R. Farwig and H. Kozono. Weak solutions of the Navier-Stokes equations with non-zero boundary values in an exterior domain satisfying the strong energy inequality. *J. Differential Equations*, 256(7):2633–2658, 2014.

[15] R. Farwig, H. Kozono, and F. Riechwald. Weak solutions of the Navier-Stokes equations with non-zero boundary values in an exterior domain. In *Mathematical analysis on the Navier-Stokes equations and related topics, past and future*, volume 35 of *GAKUTO Internat. Ser. Math. Sci. Appl.*, pages 31–52. Gakkōtosho, Tokyo, 2011.

[16] R. Farwig, H. Kozono, and H. Sohr. Global Leray-Hopf weak solutions of the Navier-Stokes equations with nonzero time-dependent boundary values. In *Parabolic problems*, volume 80 of *Progr. Nonlinear Differential Equations Appl.*, pages 211–232. Birkhäuser/Springer Basel AG, Basel, 2011.

[17] R. Farwig, H. Kozono, and D. Wegmann. Decay of non-stationary Navier-Stokes flow with nonzero Dirichlet boundary data. *Indiana Univ. Math. J.*, 66(6), 2017.

[18] R. Farwig, H. Kozono, and D. Wegmann. Existence of strong solutions and decay of turbulent solutions of Navier-Stokes flow with nonzero Dirichlet boundary data. *J. Math. Anal. Appl.*, 453(1), 2017.

[19] R. Farwig and H. Sohr. Generalized resolvent estimates for the Stokes system in bounded and unbounded domains. *J. Math. Soc. Japan*, 46(4):607–643, 1994.

[20] R. Farwig and H. Sohr. Helmholtz decomposition and Stokes resolvent system for aperture domains in L^q-spaces. *Analysis*, 16(1):1–26, 1996.

[21] D. Fujiwara and H. Morimoto. An L_r-theorem of the Helmholtz decomposition of vector fields. *J. Fac. Sci. Univ. Tokyo Sect. IA Math.*, 24(3):685–700, 1977.

[22] G. P. Galdi. *An introduction to the mathematical theory of the Navier-Stokes equations.* Springer Monographs in Mathematics. Springer, New York, second edition, 2011. Steady-state problems.

[23] M. Giga, Y. Giga, and H. Sohr. L^p estimate for abstract linear parabolic equations. *Proc. Japan Acad. Ser. A Math. Sci.*, 67(6):197–202, 1991.

[24] Y. Giga. Analyticity of the semigroup generated by the Stokes operator in L_r spaces. *Math. Z.*, 178(3):297–329, 1981.

[25] Y. Giga and H. Sohr. On the Stokes operator in exterior domains. *J. Fac. Sci. Univ. Tokyo Sect. IA Math.*, 36(1):103–130, 1989.

[26] Y. Giga and H. Sohr. Abstract L^p estimates for the Cauchy problem with applications to the Navier-Stokes equations in exterior domains. *J. Funct. Anal.*, 102(1):72–94, 1991.

[27] D. Gilbarg and N. S. Trudinger. *Elliptic partial differential equations of second order.* Classics in Mathematics. Springer-Verlag, Berlin, 2001.

[28] K. Götze. Strong solutions for the interaction of a rigid body and a viscoelastic fluid. *J. Math. Fluid Mech.*, 15(4):663–688, 2013.

[29] L. Grafakos. *Classical Fourier analysis*, volume 249 of *Graduate Texts in Mathematics*. Springer, New York, second edition, 2008.

[30] M. Haase. *The functional calculus for sectorial operators*, volume 169 of *Operator Theory: Advances and Applications*. Birkhäuser Verlag, Basel, 2006.

[31] E. Hopf. Über die Anfangswertaufgabe für die hydrodynamischen Grundgleichungen. *Math. Nachr.*, 4:213–231, 1951.

[32] A. Inoue and M. Wakimoto. On existence of solutions of the Navier-Stokes equation in a time dependent domain. *J. Fac. Sci. Univ. Tokyo Sect. IA Math.*, 24(2):303–319, 1977.

[33] H. Iwashita. L_q-L_r estimates for solutions of the nonstationary Stokes equations in an exterior domain and the Navier-Stokes initial value problems in L_q spaces. *Math. Ann.*, 285(2):265–288, 1989.

[34] N. J. Kalton and L. Weis. The H^∞-calculus and sums of closed operators. *Math. Ann.*, 321(2):319–345, 2001.

[35] J. Leray. *Étude de diverses équations intégrales non linéaires et de quelques problèmes que pose l'hydrodynamique*. NUMDAM, 1933.

[36] J. Leray. Sur le mouvement d'un liquide visqueux emplissant l'espace. *Acta Math.*, 63(1):193–248, 1934.

[37] A. Lunardi. *Interpolation theory*. Appunti. Scuola Normale Superiore di Pisa (Nuova Serie). Edizioni della Normale, Pisa, second edition, 2009.

[38] K. Masuda. Weak solutions of Navier-Stokes equations. *Tohoku Math. J. (2)*, 36(4):623–646, 1984.

[39] T. Miyakawa. On nonstationary solutions of the Navier-Stokes equations in an exterior domain. *Hiroshima Math. J.*, 12(1):115–140, 1982.

[40] T. Miyakawa and Y. Teramoto. Existence and periodicity of weak solutions of the Navier-Stokes equations in a time dependent domain. *Hiroshima Math. J.*, 12(3):513–528, 1982.

[41] A. Noll and J. Saal. H^∞-calculus for the Stokes operator on L_q-spaces. *Math. Z.*, 244(3):651–688, 2003.

[42] J. Prüss and G. Simonett. *Moving interfaces and quasilinear parabolic evolution equations*, volume 105 of *Monographs in Mathematics*. Birkhäuser/Springer, 2016.

[43] J. Prüss and H. Sohr. On operators with bounded imaginary powers in Banach spaces. *Math. Z.*, 203(3):429–452, 1990.

[44] P. F. Riechwald and K. Schumacher. A large class of solutions for the instationary Navier-Stokes system. *J. Evol. Equ.*, 9(3):593–611, 2009.

[45] J. Saal. Maximal regularity for the Stokes system on noncylindrical space-time domains. *J. Math. Soc. Japan*, 58(3):617–641, 2006.

[46] J. Saal. Strong solutions for the Navier-Stokes equations on bounded and unbounded domains with a moving boundary. In *Proceedings of the Sixth Mississippi States UBA Conference on Differential Equations and Computational Simulations*, volume 15 of *Electron. J. Differ. Equ. Conf.*, pages 365–375. Southwest Texas State Univ., San Marcos, TX, 2007.

[47] J. Serrin. The initial value problem for the Navier-Stokes equations. pages 69–98, 1963.

[48] C. G. Simader and H. Sohr. A new approach to the Helmholtz decomposition and the Neumann problem in L^q-spaces for bounded and exterior domains. In *Mathematical problems relating to the Navier-Stokes equation*, volume 11 of *Ser. Adv. Math. Appl. Sci.*, pages 1–35. World Sci. Publ., River Edge, NJ, 1992.

[49] P. E. Sobolevskiĭ. Fractional powers of coercively positive sums of operators. *Dokl. Akad. Nauk SSSR*, 225(6):1271–1274, 1975.

[50] H. Sohr. *The Navier-Stokes equations*. Modern Birkhäuser Classics. Birkhäuser/Springer Basel AG, Basel, 2001. An elementary functional analytic approach.

[51] R. Temam. *Navier-Stokes equations*. AMS Chelsea Publishing, Providence, RI, 2001. Theory and numerical analysis.

[52] Y. Teramoto. On asymptotic behavior of solution for the Navier-Stokes equations in a time dependent domain. *Math. Z.*, 186(1):29–40, 1984.

[53] H. Triebel. *Interpolation theory, function spaces, differential operators*, volume 18 of *North-Holland Mathematical Library*. North-Holland Publishing Co., Amsterdam-New York, 1978.

[54] H. Triebel. *Theory of function spaces. II*, volume 84 of *Monographs in Mathematics*. Birkhäuser Verlag, Basel, 1992.

[55] L. Weis. A new approach to maximal L_p-regularity. In *Evolution equations and their applications in physical and life sciences (Bad Herrenalb, 1998)*, volume 215 of *Lecture Notes in Pure and Appl. Math.*, pages 195–214. Dekker, New York, 2001.

[56] L. Weis. Operator-valued Fourier multiplier theorems and maximal L_p-regularity. *Math. Ann.*, 319(4):735–758, 2001.

[57] K. Yosida. *Functional analysis.* Classics in Mathematics. Springer-Verlag, Berlin, 1995.

Index

Curriculum Vitæ

05/20/89   Born in Seeheim-Jugenheim, Germany

09/04 – 06/07 **Highschool**, *Justus–Liebig-Schule*, Darmstadt, Germany, Abitur (1.8 / good)

10/08 – 04/12 **Bachelor's studies**, *TU Darmstadt*, Darmstadt, Germany, Bachelor of Science Mathematics (1.80 / good)
Bachelor's thesis: *L^p-Theory starker Lösungen elliptischer partieller Differentialgleichungen zweiter Ordnung*

04/12 – 08/13 **Master's studies**, *TU Darmstadt*, Darmstadt, Germany, Master of Science Mathematics (1.31 very good)
Master's thesis: *Eine verbesserte Energieungleichung für schwache Lösungen der Navier–Stokes-Gleichungen in allgemeinen Gebieten*

09/13 – 12/18 **Doctoral studies** *TU Darmstadt*, Darmstadt, Germany, research assistant in the working group *Partielle Differentialgleichungen*

10/18/18   **Submission of the doctor's thesis (Dissertation)**
The Stokes and Navier–Stokes Equations in Exterior Domains: Moving Domains and Decay Properties at *TU Darmstadt*, Darmstadt, Germany

12/07/18   **Defense of the doctor's thesis**